The Last Giants

The Rise and Fall of the African Elephant

Levison Wood

HODDER

First published in Great Britain in 2020 by Hodder & Stoughton
An Hachette UK company

This paperback edition published in 2021

1

A CIP catalogue record for this title is available from the British Library

Paperback ISBN 9781529381160
Hardback ISBN 9781529381122
eBook ISBN 9781529381146

Typeset in Bembo by Hewer Text UK Ltd, Edinburgh
Printed and bound in Great Britain by Clays Ltd, Elcograf S.p.A.

Hodder & Stoughton policy is to use papers that are natural, renewable
and recyclable products and made from wood grown in sustainable
forests. The logging and manufacturing processes are expected to
conform to the environmental regulations of the country of origin.

Hodder & Stoughton Ltd
Carmelite House
50 Victoria Embankment
London EC4Y 0DZ

www.hodder.co.uk

For all the wildlife rangers and conservationists
who dedicate their lives to protecting all species

Levison Wood is an award-winning author, explorer and photographer who specialises in documenting people and cultures in remote regions and post-conflict zones. His work has taken him around the world leading expeditions on five continents and he is an elected fellow of both the Royal Geographical Society and the Explorers Club.

Levison's second book, *Walking the Himalayas*, was voted Adventure Travel Book of the Year at the 2016 Edward Stanford Travel Writing Awards and his other books, *Walking the Nile* and *Walking the Americas*, were both *Sunday Times* bestsellers. He has presented several critically acclaimed documentaries including *From Russia to Iran: Crossing the Wild Frontier* where he re-traced part of his Silk Road adventures in a four-part series for Channel 4.

Also by Levison Wood

Walking the Nile
Walking the Himalayas
Walking the Americas
Eastern Horizons
Arabia
Incredible Journeys

With special thanks to Dr Lucy Bates, Visiting Fellow at the School of Psychology, University of Sussex and Dr Graeme Shannon, Lecturer in Zoology in the School of Natural Sciences, Bangor University without whom this book would not have been possible. Their research and contribution has been invaluable, and their work instrumental in how we understand the world of elephants.

Contents

Introduction

Once, when I was a young boy, my father took me to an art exhibition during the school summer holidays. At the time, my dad was a keen amateur painter, one of his many changing hobbies, and a famous artist called David Shepherd had brought his paintings to the local town of Leek where they were on display. What's more, Mr Shepherd was in town himself, signing books and talking about his pictures.

Now at the age of eleven, I can't say I knew much about art, but I went along to humour my dad, who really wanted to meet this great man. When I got there, much to my relief, I found that the pictures were very good. There were lots of paintings of trains, planes and important people, but the ones I liked the most were the pictures of animals. There were tigers, zebras and rhinos – although the paintings that intrigued me the most were the ones of elephants.

'Do you like elephants?' a voice called out from behind me, as I was staring up at the vast canvas.

I turned around to be confronted by a scruffy-looking, white-haired man, who appeared to me to be very old. I told him that I had never seen an elephant before in real life, but I'd read about them at school and I'd seen them on the David Attenborough documentaries. 'Well, one day I'm sure you'll see them for yourself, in Africa perhaps,' he said, with a patient smile. He put out his hand and I shook it. It was David Shepherd, the artist himself.

'Ask Mr Shepherd a question,' my dad insisted. My mind went blank for a moment, before it occurred to me to ask whether or not he had always been a good artist. Mr Shepherd stroked his chin and smiled.

'Young man,' he said, 'shall I show you one of my first efforts at painting?'

I nodded.

David Shepherd turned around and motioned for me to follow him to the corner of the room, where he had some bags and a large plastic folder, which he picked up and opened. He rustled around and out of it he pulled a yellowed piece of paper, no bigger than a normal A4 sheet. He handed it to me. I looked down and my astonishment must have been quite apparent.

'Not very good is it?' he said, beaming. I didn't know what to say. My dad had always taught me to be polite, but there was no hiding the fact that the sketch of some seagulls was in fact pretty bad. I shrugged and looked at the floor in embarrassment.

'Don't be shy, young man. It's terrible. But you know what? I put my mind to it and spent all my time practising until I became good enough that people wanted to buy my pictures, and then I could call myself an artist.'

I looked at the seagulls again. I was pretty sure I could do better than that myself, even at my age, and decided there and then that I wanted to become an artist too, and see for myself the wild elephants in Africa.

A year or so later, I found myself in the steamy coastal rainforests of southern Kenya, on holiday with my parents, surrounded by tall trees filled with glinting fish eagles and bewitching grey parrots. In the middle of the jungle lay a wooden treehouse made of cedar, which jutted into the canopy. Looking down from its beams in the half light of dusk, I could see the murky pools of Shimba Hills watering hole reflecting the tropical yellow moonlight.

The erupting orchestra of bullfrogs and cicadas sang a melody of exotic brilliance across the jungle and a magical scene began to unfold. There was movement below. Shapes teased the eye as blackened, boulder-like forms shifted through the foliage; huge yet silent ghosts seemed to float across the forest floor, gathering at the water's edge.

Elephants, dozens of them, appeared as if out of nowhere on their nightly pilgrimage to an ancient shrine. To the eyes of a child, it was wondrous and enchanting, and I stood transfixed – my first glimpse of these magical beasts in the wild. I knew they could never be my last. It was the beginning of a lifelong love affair with Africa and its indigenous creatures.

Since then, although I never became an artist, I have travelled the length and breadth of the continent in various guises, and whenever I've had the chance, I've tried to make time to meet elephants. I've been fortunate enough to go on safari in wonderful and exciting countries such as South Africa, Zambia and Zimbabwe, and to trek through wilderness areas and national parks as far afield as the Congo and Malawi.

Over the course of nine months between 2013 and 2014, I walked the length of the great Nile River, from Rwanda to Egypt, hiking over 4,000 miles and witnessing elephants in their natural habitat in Tanzania, Uganda and South Sudan, where I was lucky enough to be invited by the conservation charity The

Tusk Trust to see their organisation's work in protecting this species up close and personal on the ground.

Then again in the summer of 2019, I spent a month in Botswana walking with elephants on their annual migration towards the Okavango Delta, which gave me a great opportunity to see some of the very complex problems facing both local people and conservationists who strive to protect elephants.

As the twenty-first century progresses into its third decade, elephants are regarded as an endangered species. In my lifetime, the elephant population in Africa has halved from around a million in 1982 to only 415,000 in 2019. Between 20,000–30,000 elephants each year are killed as a result of poaching and the illegal trade in wildlife. That's one elephant slaughtered every twenty minutes. Many more are forced away from their traditional feeding grounds because of encroachment by humans onto wilderness areas, changes in land use, and the ever-greedy market for ivory and animal parts.

Like most people, I find the statistics horrifying, but have tried as much as possible to keep an objective standpoint. I am not an expert in elephant biology, psychology or conservation. I merely profess a deep interest and I hope this book will appeal to those of a similar mindset. Of course, I am limited in scope as to what I can hope to achieve. There are many other books out there by academics and scientists who have spent a lifetime in the field and go into far more detail, and I have included a selected reading list for those who want to learn more.

This book gives an outline of where elephants came from; their evolutionary past, and their place in ecology. It examines the inner and outer workings of an elephant, looking at their biology, their psychology (insomuch as our limited understanding will allow) and how they impact their own environment through feeding and migration. I try to show how the long life and sociality of elephants is key to their success and survival, and yet might also be the foundations of their demise.

After that I explore what impact we as humans have had on elephants, in terms of the ivory trade, hunting and poaching, as well as changes in land use across Africa. In doing so, I hope to summarise how we have allowed elephant numbers to plummet and the influence recent human history has had on the species – in particular colonialism and its aftermath – which has undoubtedly had a major effect on all African wildlife. The policies and prejudices that we are dealing with now all have roots in decisions that were made a hundred years ago.

Finally, I try to forecast the future, in terms of what the world would be like without elephants, and also, on a happier note, how we might be able to coexist with this noble animal. After all, the future is not yet written.

What we do in the next few years will determine the next few thousand years.

Sir David Attenborough's words will no doubt ring true to many of us as we peer over the abyss at the end of the Holocene. Let's hope we all make the right decisions. I hope that you will find this book an introductory glimpse into the lives of Africa's elephants, and that you will go on to play your own part in helping to save them.

We owe it not only to the elephants, but to our planet and ourselves to do what we can to preserve the last giants.

London
October 2019

I

A Brief History of the African Elephant

The path was littered with dead branches and twigs, and the skeletal spines of acacia scrub poked into our sides like needles. Every step forward had to be done with the utmost care not to make a sound, and we crept forward like hunters sneaking towards their prey. Kane, my local bushman guide, led the way, his rusty old spear pointing forward in the direction of our quarry. I watched as he delicately tiptoed over the litter of foliage as quiet as a mouse. I tried my best to follow in his footsteps, but his stride, though silent, was fast and deliberate.

'Keep up, and be quiet,' he halted briefly and whispered, staring intently into my eyes with a passion I hadn't seen before. 'One noise, one false move and they'll trample you to pieces.'

I nodded without a peep and looked around. I couldn't see anything except the surrounding trees. We were in the middle of a dense thicket of palms and thorny undergrowth, trying our best to get towards a cluster of baobab trees where the herd were browsing.

'This way, *shhh*,' whispered Kane. He held his hand up motioning for me to move. But I was half-balanced on one leg, and before I could take another step, I stumbled and put a foot straight down on a twig that snapped with a clear, crisp crack.

Kane whipped his head around and grimaced. '*Shhhh!*' putting a finger to his lips and screwing up his face, which made him look like an angry warthog.

I pursed my lips and shrugged. I couldn't even see where the herd was.

'Let's get closer,' he said. 'But be quiet.'

Closer we got, padding forward until I could hear the rustle of bushes up ahead. 'There!'

Kane pointed into a small clearing at the base of the fat baobab tree. A huge bull elephant was ripping a branch to shreds with his trunk and feeding the mulch into his mouth. There was another crunch to my right and I looked over. Not twenty feet away was another bull, even bigger that the first, except this one wasn't eating. He had his trunk waving around in the air pointing in our direction.

'We call him a sniffer dog,' said Gareth, who'd been trailing behind me. Gareth was a professional hunter and was keeping watch to the rear, gripping with both hands the bolt-action rifle that was loaded with high-calibre ammunition. 'He sniffs out the air for danger while the rest of the boys eat.'

'Has he seen us?'

'They have bad eyes,' interrupted Kane, 'but he knows we're here for sure.'

'Come on, see that fallen tree up ahead, let's get there.'

We darted forward, as quickly as we could without breaking into a run, me following Kane, with Gareth behind. Never run, never run, never run. It had been drilled into me by Gareth before we set off. An elephant can run at twenty-five miles an hour, far outpacing any human.

'Duck there,' said Kane. 'If he charges, we'll be safe if you bury yourself under the log.' I did as I was told, crouching down by the log. I didn't fancy my chances, though; if the bull came at us, the tusks on the elephant could surely rip it apart in no time.

'Don't worry,' said Gareth. 'I'll tell you if we need to run.'

'But you said never run,' I protested.

He shrugged. 'Look, when I say never, I mean sometimes you don't really have a choice. Usually an elephant will do only a mock charge, unless he's really pissed off. Or if he's been shot at, of course. Then he means business, especially if he sees my rifle.'

I thought back to my own close shaves, such as the time in Malawi when I'd been charged by a massive female elephant on the Shire river and my local guide had needed to fire a warning shot towards the rampaging beast. Then there was the time in Uganda, when a whole herd of elephants wandered straight through my camp at night, almost squashing me in my tent.

I remembered the story of a fellow paratrooper, who'd been gored by an elephant in the wilds of Kenya – ripping his arm in two – and how, a couple of months before I set off to Botswana, another soldier in the British army had been killed by an elephant whilst on an anti-poaching patrol. There was no doubting that elephants are dangerous wild animals, whose relationship with humans is, at best, turbulent.

So what on earth was I doing, travelling on foot through some of the most dangerous terrain in Africa, trying to research more about them?

It was a good question, and there'd been plenty of times when I'd been photographing them that I'd been forced to question my own sanity, but I always calmed myself with the thought that, in spite of their massive size and potential for causing damage, they were also highly intelligent, gentle beasts that were capable of great compassion, and needed to be understood.

We sat still, watching as more males arrived, grazing on the low-lying branches, seemingly unaware of our presence, apart from the 'sniffer dog', who never stopped wafting his trunk in our direction.

'Right, I think it's time to go,' said Gareth, calmly. 'There's about ten of them, and if any more come we might find ourselves surrounded, and that would end badly.'

I agreed. We'd got very close, and I'd been lucky to get some great photographs and observe the herd up close and personal, but I didn't want to push my luck.

As we tiptoed backwards, I noticed movement in the bushes right ahead. It was the 'sniffer dog' again, and he'd started to follow us; slowly at first, but he seemed determined not to lose us. Anyone not acquainted with elephant behaviour might have thought he was merely curious, but Gareth reminded me of the urgency.

'Pick up the pace, Wood, get moving. He wants to let us know that he's the boss.'

Kane led the way, jabbing his spear into the bushes to clear a way. 'Faster, he's coming.'

I turned around to see the young bull gaining on us.

'Okay, move now!' shouted Gareth, and this time there was no doubting the urgency in his voice. At the same time Gareth cocked his weapon and I shuddered at the familiar sound of metal clunking and hoped beyond anything that he wasn't forced to use it. I picked up my pace and started to jog, checking over my shoulder every few paces.

Suddenly I heard the violent snort of the bull as he crashed through the thicket, at which point he couldn't have been more than twenty feet away. There was a loud trumpet as the bull smashed against the side of a tree and the thud seemed to vibrate the earth.

Now he began to run properly, straight towards us.

'Go, go, go!' Kane pointed his spear towards the edge of the treeline, where a gnarled uprooted tree blocked the path. 'Jump!' he shouted, and with all my energy I launched myself over the natural barrier into the clearing beyond. Kane, who'd done the same, landed with a thump next to me, and meanwhile Gareth had the good sense to run around the side.

The rampaging bull skidded to a halt in front of us, violently shaking his head and screaming the most terrible noise, which seemed to split the atmosphere of the forest in two. He stamped his feet and waved his ears in a show of ferocious terror. Then with one final snort and whip of his trunk, he simply turned around and plodded away.

Gareth was still catching his breath, and I could feel my heart beating in my chest and the adrenaline searing through my gut. That was a close call.

Kane burst out laughing and shook his head. 'Well, he was a show-off, wasn't he?'

When we look at elephants, it is often through a photographer's lens, or from the comfort of a safari vehicle, gazing at them through a pair of binoculars. It can sometimes feel voyeuristic and surreal. A caged human in an animal's world, a sort of zoo reversed. Yet when I was walking in the footsteps of the herds, treading in the wake of their destruction, vulnerable and ever alert, nothing could have felt more natural.

There's something exhilarating about being at the mercy of nature in its rawest form, of putting yourself into the mind of a wild animal. Perhaps it is some primal emotion taking us back to our prehistoric roots, when human and giants roamed together in constant communion, fear and understanding; back to a time of pure survival, when it was essential for us to know intimately the ways of the beasts.

Elephants have been around for far longer than human beings, and all throughout our own evolution and history we have been in their company on the plains and in the forests until very recently, all around the world. But before we go on to look at where these creatures came from, it's important to think about why they are important, and how our relationship has intertwined.

You may have heard the parable of the elephant and the blind men. It tells a cautionary tale about six blind men who encountered this strange animal and decided they must learn what it was like by touching it. Each blind man felt a part of the elephant's body, but only one part, such as its legs or ear or tusk. They then had to describe the beast to the audience based on their limited experience. Their descriptions of the elephant were, of course, wildly different from each other.

The first man, whose hand landed on the trunk, quite naturally remarked, 'This being is like a thick snake.' Another one, whose hand reached its ear, said it seemed like a kind of fan. As for the third person, whose hand was on its leg, he thought the elephant was a fat pillar, like a tree trunk. The blind man who placed his hand upon its side believed the animal was 'a wall'. Another, who handled its tail, described it as a rope. The last one stroked its tusk, claiming that the elephant was hard, smooth and pointy, like a spear.

In some versions of the story, the blind commentators each suspect that the others are being dishonest and they come to blows. The moral of the parable is that humans have a tendency to claim absolute truth based on their own limited, subjective understanding and are prone to ignoring other people's experiences, even though they may be equally true.

The nineteenth-century poet John Godfrey Saxe wrote:

It was six men of Indostan
To learning much inclined,
Who went to see the Elephant
(Though all of them were blind),
That each by observation
Might satisfy his mind

And so these men of Indostan
Disputed loud and long,
Each in his own opinion
Exceeding stiff and strong,
Though each was partly in the right
And all were in the wrong!

He concludes:

So oft in theologic wars,
The disputants, I ween,
Rail on in utter ignorance
Of what each other mean,
And prate about an Elephant
Not one of them has seen!

In the modern era of polarised politics, antagonistic populism and fake news, perhaps it's worth taking a moment to learn something from the humble elephant.

When I was little, I was treated to my own particular version of the parable, when I used to visit my Grandad Curzon. The 'elephant graveyard' became my favourite childhood game. It was a particularly gruesome bit of child's play that involved my grandfather blindfolding me and telling me the story of a blind explorer who got lost in the jungles of the Congo, sightless perhaps after having caught malaria or some other nasty tropical disease.

I would revel in excitement as he walked me hand in hand around the garden, and through the 'jungle' (rhododendron hedge), past the lethal acacia (rose bushes) and taking care not to wake up the sleeping hyena (the neighbour's dog). When we passed through the caves of doom (the porch), I knew we were almost reaching the secret destination of our mission, because, despite my blindness, I could feel the warmth of the volcano (the hearth fire).

This was the infamous elephant graveyard, where all the African elephants go to die. It was here that I'd be put through a series of ordeals to test my manhood. I would hold out my hand and be guided by my grandfather to reach out and grasp an inanimate object and have to guess which part of the rotting elephant carcass it was. There were the bones (rack of lamb), the eyeballs (a squishy tomato), the guts (long party balloons), brains (a wet sponge), teeth (his false teeth), and of course, the tusks, in the form of a sharpened cucumber.

If I guessed correctly, then I was able to navigate my way out of the dreaded place, claim my treasure from the grotto (a shiny new 50 pence piece), and regain my sight before bedtime ...

Elephants have existed in our collective consciousness for as long as we humans have been roaming the African plains and living alongside them. Even in the cities and towns of Europe and the Americas, where elephants were hunted to extinction long ago, the beasts still survive in the form of hearsay, myths and legends.

In many African cultures, the elephant is revered as a creature that embodies the human virtues of intelligence, wisdom and physical strength. The Kamba tribe in Kenya believe that elephants were once human beings. As an old myth goes, there was a poor fellow who set off to find a wealthy and generous man known for his wisdom. The poor man desperately wanted to discover the secret to being rich. After a long journey, the poor man arrived at a beautiful house surrounded by fertile pastures with abundant herds of cattle and sheep.

Here, the wise and rich man generously offered the poor man a hundred sheep and a hundred cows, but the poor man refused, demanding not charity, but the rich man's secret to success. So instead, the rich man gave the poor man an ointment and told him to rub it on his wife's front teeth.

The poor man left and somehow convinced his wife to participate, because it would make them very wealthy. Soon after, her teeth began to grow and grow and toughened into ivory tusks the length of a man's arm. On seeing the incredible spectacle, the poor man imagined untold riches and pulled the tusks out of his wife's mouth and sold them for a lot of money.

After that, emboldened and excited, he began rubbing the ointment on his wife's teeth again. But this time when his wife grew tusks, she understandably refused to let her husband touch them, and then, before either of them knew what was happening, her entire body started to change. She became fatter and fatter, and her skin became wrinkled and baggy and grey, and, as if to add insult to injury, her nose got longer and longer until she became a fully fledged elephant. Her husband was so alarmed by her that she ran away deep into the forest where, after a period of lonely sadness, she gave birth to their sons and daughters – the first line of elephants.

Elephants and humans have coexisted in Africa for hundreds of thousands of years, and elephant wisdom is seen as sacred. In Gabon in West Africa, the three great animal chiefs are the leopard, deemed powerful and cunning; the monkey – malicious and agile; and of course, the elephant – wise and strong. People from Ghana and Sierra Leone regarded elephants as past human chiefs and deceased ancestors. One Zulu legend from South Africa tells of a young girl, outcast from her tribe, who in her wanderings, finds a kind and hospitable elephant and marries him; their children, who benefitted from the magic of the beasts, in turn gave birth to a line of powerful chiefs and eventually the forerunners to the royal family.

But not all myths are so reverent: in Namibia, the story goes that the elephant got its long trunk because it was so slow and clumsy that it couldn't fend off a crocodile, which bit the docile animal and stretched its nose into the now trademark trunk.

In the Congo, legends have persisted for centuries that witchcraft can turn people into animals, and even today many villagers will blame their neighbours when elephants run amok and raid crops or kill people, saying that it was down to voodoo or magic, which is often a convenient way to bad-mouth an enemy or have an excuse to plunder a nearby village.

Whether good or bad though, elephants have featured heavily in cultural symbolism, art and storytelling around the world since ancient times. I remember being rather surprised when walking across the Sahara Desert in Sudan to discover prehistoric etchings on a rocky outcrop depicting all sorts of animals, including elephants. They date back thousands of years to a time when North Africa was a lush, green savannah, like much of the rest of Africa. In the Tadrart mountains, on the border of Libya and Algeria, lies some of the best-preserved rock art in Africa, including a remarkably well-proportioned picture of an elephant that's 12,000 years old.

North African elephants have been extinct for 1,500 years, but their legacy lives on in Mediterranean culture. The Biblical behemoth that features in the Book of Job is described as a fantastical monster, which sounds suspiciously like a pachyderm:

Behold now behemoth, which I made with thee; he eateth grass as an ox.

Lo now, his strength is in his loins, and his force is in the navel of his belly.

He moveth his tail like a cedar: the sinews of his stones are wrapped together.

His bones are as strong pieces of brass; his bones are like bars of iron.

He is the chief of the ways of God: he that made him can make his sword to approach unto him.

Surely the mountains bring him forth food, where all the beasts of the field play.

He lieth under the shady trees, in the covert of the reed, and fens.

The shady trees cover him with their shadow; the willows of the brook compass him about.

Behold, he drinketh up a river, and hasteth not: he trusteth that he can draw up Jordan into his mouth.

He taketh it with his eyes: his nose pierceth through snares.

In Ancient Egypt, elephants were prized as both war machines and status symbols, dead and alive. Elephants featured in hieroglyphics as a testament to a time when their range was on a global scale.

In Ancient Greece, too, the elephant found its way into popular culture after Alexander the Great encountered war elephants on his travels to India, and subsequently incorporated them into his own army. When the ancient sailors dug up the skulls of prehistoric dwarf elephants on the island of Cyprus, it's easy to forgive them for thinking that they must have belonged to the remains of the mythical one-eyed Cyclops.

Then, of course, Ancient Rome had its own encounters with the beasts when the North African leader Hannibal brought an army of war elephants halfway across Europe, and famously over the Alps, to invade Italy. At least one elephant was used in Caesar's invasion of Ancient Britain, 'which was equipped with armour and carried archers and slingers in its tower. When this unknown creature entered the river, the Britons and their horses fled and the Roman army crossed over.'

Over the centuries, elephants have variously been revered, feared and worshipped – their image being used to symbolise all that is great and powerful.

In 1255, the King of France gave an elephant to the English monarch King Henry III as a unique gift. It was kept in the gardens of the Tower of London, and medieval Londoners flocked to see the mysterious beast. While confined to the lawns of the metropolitan fortress, it was said that the elephant was fed prime cuts of beef and rather enjoyed a bucket of red wine. It's no wonder he is reputed to have died from obesity. Nowadays the tower hosts a sculpture of the poor creature, peering down from its haunted walls.

Napoleon commissioned an artist to design an elephant monument to be built outside the Bastille. It was meant to be an enormous bronze sculpture demonstrating the emperor's power in Africa, but it never got past the plaster-cast model stage, which ended up being abandoned, and the project eventually came to symbolise futility and folly in Victor Hugo's *Les Misérables*.

Many African nations, including South Africa, use elephant tusks in their coats of arms to represent wisdom, strength, moderation and eternity. The elephant is symbolically important to the nation of Ivory Coast, whose heraldry features an elephant head as its focal point, and in the western African Kingdom of Dahomey (now part of Benin), the elephant was associated with the nineteenth-century rulers of the Fon people, whose flag depicted an elephant wearing a royal crown.

In Denmark, there is a chivalric order called 'Order of the Elephant', which is the country's highest honour, usually bestowed only upon monarchs and heads of state. Indeed, Her Majesty Queen Elizabeth II is herself a 'Knight of the Elephant'.

Even across the pond, the elephant is still visible in everyday politics, having become the symbol of the Republican Party – a throwback to *Aesop's Fables* and the story of the Rat and the Elephant.

Elephants feature in our thoughts and our language. If someone says, 'Think of a big grey animal,' you are most likely to think of an elephant, rather than a hippo or rhino, wolf, tapir, gorilla or seal. Elephants have become their own idioms: a 'white elephant' being a burdensome thing that's difficult to get rid of; whilst an 'elephant in the room' is the inconvenient truth that nobody wants to speak about.

Elephants have been written about, painted, mocked and allegorised for millennia. Elephants have featured alternately as a symbol of natural might, of fearsome magnitude, peaceful coexistence and utilitarian commercialism.

They have been background extras and leading characters in books for over 2,000 years. Elephants represent nature at its biggest: they symbolise wilderness but also danger, fear as well as courage; they personify war and peace; brute force and the height of intellect. Children love elephants, and adults love them too, because elephants, for as long as we can remember, represent us.

Over 100 million years have passed since the common ancestor of humans and elephants – a small, shrew-like animal – walked the earth. We diverged at a time when dinosaurs still ruled the world, yet we maintain a fascination with elephants that is hard to define, our fates seemingly intertwined throughout history.

From their huge size and strange appearance to their extraordinary senses and incredible brains, it appears that everything about them is unusual, extreme, or unique, and yet in so many ways they are more like us than we would care to admit. They are without a doubt one of the most remarkable and fascinating creatures on earth.

2

Ancestors and Evolution

Looking at their body size, where they live, and the kind of environments that they live in, it would be easy to assume that the closest living relatives to elephants would be other megaherbivores, such as rhinos and hippos. But in fact, genetic analysis has revealed that the closest living relative of the elephant is the rock hyrax – a furry, rodent-like creature that looks a bit like a guinea pig and isn't much bigger.

Elephants, along with hyrax, and, believe it or not, the aquatic manatees and dugongs, belong to the branch of mammals known as Afrotheria, which simply means 'African beasts'. This ancient group split off from other mammals at a time when Africa was an island continent, probably during the Cretaceous geological period, tens of millions of years before Triceratops and Tyrannosaurus dinosaurs were thundering around the American plains!

Given the long evolutionary history of the Afrotheria, surprisingly few living mammals fall into this group, and those that do are exceptionally diverse. As well as the elephants, hyraxes, and manatees, Afrotheria also includes elephant shrews, golden moles, and tenrecs, which are shrew and hedgehog-like mammals that are mainly found on Madagascar. The group is unique in that is contains one of the smallest living mammals, the long-eared tenrec, which weighs just 5 grams – not much more than a penny coin – as well as the largest, the African savannah elephant. It was only thanks to scientific developments in genetics over the past twenty-five years that we had any idea that these animals were in one related group.

While rock hyrax and the other little creatures are native to the hills, plains and valleys of Africa and the Middle East, manatees and dugongs float around in the tropical waters of Central and South America and off the warm coasts of the Caribbean, Indian and Pacific Oceans. Yet, despite their obvious differences, these seas creatures share a number of traits with elephants: manatees are the only herbivorous marine mammal; they have four small nails at the end of each flipper that are very similar to the toenails of an elephant; and they also have a prehensile upper lip that they use to grasp hold of marine vegetation, much like a stubby trunk. They even have similar teeth, with incisors that resemble tusks, as well as the horizontal molar teeth displacement that also occurs in elephants.

Hyraxes have flattened nails, rather than the claws found on most similar-sized land mammals. They even have small tusks too that develop from incisors. And, like elephants and manatees, their mammary glands (which produce milk for their babies) are near their front legs. In all other mammals, except primates, milk teats are found between the rear legs. A final trait shared with elephants is that their testicles stay inside the abdomen, rather than swinging around like a monkey, bull or human. So, despite the aeons since the common ancestor of these species was alive in the swamps of North Africa, there are still plenty of visible clues to their shared heritage.

Fossil evidence helps us to work out when the anatomical features of living elephants evolved, and why. If we were to journey back over 35 million years ago to Northern Africa, we might be fortunate enough to see a pig-sized animal that looked somewhat like a modern-day tapir, foraging in the soft vegetation around rivers and lakes. This was *Moeritherium*, one of the earliest proboscideans – the group within Afrotheria that specifically contains elephants and their relatives.

Moeritherium died out without leaving any descendants, so it is not a direct ancestor of today's elephant species, but they shared features with other proboscideans, like a flexible upper lip, which, like a modern trunk, was used for grasping and handling food. They also had short tusks, although these were more tooth-like than the tusks of a modern-day elephant.

Part of the skull and teeth of another very old proboscidean – *Eritherium* – was recently discovered in Morocco. It is dated to 60 million years ago, which was a time of rapid evolution and change for mammals following the demise of the dinosaurs. *Eritherium* was tiny, standing only 20 cm high, no bigger than a well-fed domestic cat, making it a thousand times lighter than a modern African elephant. But, despite appearances, it was its unique teeth that allowed scientists to identify it as the earliest ancestor of the elephant!

Other than humans, the evolutionary history of proboscideans is one of the best-mapped mammalian lineages, with around 175 species identified. They are divided up into five main groups, which are referred to as 'super-families' by biologists.

There are the early proboscideans (such as *Eritherium* and *Moeritherium*), the deinotheres, mastadons, gomphotheres, and elephantidae. Taxonomy is a complex business, though, particularly when dealing with creatures that have been extinct for millions of years and have left only a few fossilised fragments behind to help us. Every time a new fossil or thread of genetic evidence becomes available, scientists end up adjusting relationships between super-families and updating estimated ages, and we're still a long way off from knowing the full story.

However, from the fossils we do have, we can see that the proboscideans were remarkably successful, having lived on every continent except Antarctica and Australia in their 60 million years on earth, in environments as diverse as deserts, tropical forests, mountain ranges and the Arctic tundra.

Around 20–30 million years ago, the deinotheres ('terrible beasts') appeared on the scene. Instead of having tusks in the upper jaw, like elephants, they had downward-curving tusks looping down from their lower jaw. They started small, but some grew quickly (in evolutionary terms) reaching an impressive shoulder height of four metres.

Deinotheres stuck around for about 20 million years; the equivalent group of our human ancestors has been around for only two million, and our species, *Homo sapiens*, only 300,000 years. Deinotheres were so successful because of their increasing body size, meaning they could tolerate lower quality diets such as fibrous, hard-to-digest plants, which allowed them plenty of flexibility because they could graze more widely.

Around 25 million years ago, the climate began to warm, and big, open grasslands and tundra appeared across the world. Mastodons and the fantastically named gomphotheres marched out from Africa, across Eurasia, and into this favourable new environment in North America, with some species – such as the American mastodon – dominating the landscape until around 10,000 years ago.

Like modern-day elephants, mastodons had tusks that emerged from the upper jaw, and the thick enamel and ridging on their teeth shows that they predominantly browsed on woody plants. The largest known of all these creatures, a mastodon called *Mammut borsoni*, was well over four metres tall and reached the colossal weight of 18 tonnes. This puts it in the same league as the hornless rhino, *Paraceratherium*, a monstrous beast that lived across Eurasia 34–20 million years ago and has been officially recorded as the largest mammal species that ever lived.

However, there is some tantalising evidence that a third species may outstrip both these giants. Partial leg bones discovered in the 1800s, from a straight-tusked Asian elephant group called *Palaeoloxodon*, suggest an animal with a height of over five metres and a body weight of 22 tonnes. They would make today's elephants look like small fry.

The gomphotheres are the most diverse group of the proboscideans, first appearing in Africa 24 million years ago, then spreading across the globe, before mostly becoming extinct 11,000 years ago. Many gomphothere species had four tusks and while some had trunks that are similar to modern elephants, others had shorter snouts more reminiscent of today's tapir, which is more closely related to rhinos and horses than it is to elephants.

This brings us to the *Elephantidae* – the group that includes the three remaining species of proboscidean alive today, as well as the extinct mammoths (*Mammuthus*) and *Palaeoloxodons*. *Elephantidae* emerged in Africa in the late Miocene period, somewhere between 6–8 million years ago. This was truly a time of trunks, when deinotheres, gomphotheres and mastodons all ranged far across the planet from the tropical forests of America to the arid grasslands of Africa and Asia.

66MYA

PALEOCENE

ERITHERIUM

56MYA

EOCENE

MOERITHERIUM

PALEOMASTODON

33.9MYA

OLIGOCENE

DEINOTHERIUM

23.03MYA

GOMPHOTHERIUM

MIOCENE

PLATYBELODON GRANGERI

EUBELODON

5.33MYA

PLIOCENE

2.58MYA

STEGOMASTODON

MAMMUT

PLEISTOCENE

11'700YA

HOLOCENE

MAMMUTHUS   ASIAN ELEPHANT   AFRICAN ELEPHANT   FOREST ELEPHANT   PRESENT

We also know a little bit about the behaviour of these early elephants. Scientists recently used aircraft to study the ancient fossilised trackways of an unknown proboscidean species in the deserts of Eastern Arabia, near to modern-day Dubai. Seven million years ago, we know that a single group of around thirteen animals, of different ages and sizes, had moved across the muddy ground in a coordinated fashion, much like elephant families move across the savannah today.

There was also the single track of a large individual, suggesting that the society of these ancient creatures was sexually segregated like today's elephants, with independent males usually walking separately from the family group. The social patterns seen in today's elephants stretch back millions of years into evolutionary history.

Around 5 million years ago there was another enormous transition on our planet as the earth became cooler and dryer. The forests began to shrink further, while deserts and grasslands got bigger. As a result, the availability of good quality food for large herbivores declined and those that could not adapt went extinct. Large body size (and perhaps large brain size) was an advantage, because bigger animals can eat more low-quality food to get their daily nutrients and energy.

Then, about 2 million years ago at the start of the Pleistocene epoch, another period of rapid climate change saw the earth cool even further. Elephants were forced to adapt yet again. Glaciation meant that some species became isolated and others were squeezed into areas that were already being inhabited by other animals. Fluctuating sea levels even led to some elephants becoming trapped on islands.

Being large is typically unnecessary on islands without predators, because there is no need to waste time and energy growing big to avoid being eaten, when there is no one to eat you. So, many island elephants underwent rapid and radical shrinking (rapid in evolutionary terms, at least). One such species, the

ERITHERIUM

PALEOMASTODON

GOMPHOTHERIUM

MAMMUTHUS

AFRICAN ELEPHANT

Cyprus dwarf elephant – a straight-tusked species that became extinct around 13,000 years ago – weighed only 200 kg and was the size of a large dog.

Despite their huge success, with a considerable evolutionary history and the fact that they could be found almost all over the world, there was a sudden and dramatic extinction of proboscideans at the end of the Pleistocene, between 15,000 to 10,000 years ago. Mastodons, gomphotheres and mammoths all died off at an unprecedented rate. Only one small population of woolly mammoths survived much longer. They remained isolated on the Wrangel Island of Siberia until 4,000 years ago, which is remarkable, because by that time the pyramids of Egypt had already been built!

It wasn't only proboscideans who suffered during this period. Thirty-five groups of large mammal became extinct in North America alone, including the giant sloths, sabre-toothed tiger and the American cheetah. Before these extinctions, the diversity of mammal species in the Americas exceeded that of modern-day Africa.

Two main ideas have been put forward to explain these sudden declines. The first suggests that, as had happened before, rapid climate change altered vegetation to such an extent that large herbivores and their predators could not adapt quickly enough and died out from starvation. The second, equally grim hypothesis suggests that this period saw the emergence of a super-predator, which used its superior intelligence and employment of tools to kill off the big beasts. That predator was, of course, us: modern humans armed with Stone Age technology.

Increasingly, the evidence points an accusatory finger towards us. Either way it's hard to brush off as coincidence that this cascade of extinctions took place over a remarkably short time period – a few thousand years. As the ancestors of modern humans evolved in Africa and migrated in successive waves across Eurasia, large mammals would have experienced varying levels of exposure to this new predator, depending upon where they lived.

African mammals co-evolved with hominids (the great apes, which includes us) and were slightly better adapted to dealing with our ancestors. Better hearing, smell, and an innate fear of people, perhaps. Species in Australia and the New World, however, were caught completely unawares by our sudden arrival, and were much more vulnerable.

When we look at changes in the distribution of plants and animals over time, the large mammal extinctions correlate almost perfectly with the arrival of modern humans. Those species that had the least time to adapt to the arrival of man were the most likely to die out.

There is even some evidence that mammoths were the preferred targets of early human hunters. It makes sense: killing one of these enormous beasts would provide a huge amount of meat for a hunter-gatherer family, which made the effort and risk of hunting them worthwhile. But, as prehistoric cave art depicting them shows, mammoths provided more than simply meat. Ivory is a much better material than deer antler for making the spear and arrow points needed for big-game hunting, so the ivory of slain mammoths was used to hunt more effectively and kill even more mammoths and other megafauna.

Climate change almost certainly played some role in driving a number of extinctions, but this is increasingly believed to be a secondary factor in the collapse of large mammals. South America suffered some of the greatest losses, yet the climate there was far more stable than in Africa, which experienced comparatively few megafauna extinctions over the same period.

Many of the large herbivore species that went extinct were

generalist foragers, who could eat grass or browse for food depending on availability. All the evidence suggests that there were still plentiful foraging opportunities, and some species, such as the American mastodon, went extinct even though their preferred plant species have remained abundant up until present day. The only viable culprit for these mass extinctions appears to be the human race.

Human pressure on animal populations. Man's desire for ivory. Climate change. This all sounds eerily familiar. We – or at least our ancestors – have to hold up our hands for the dramatic decline in the number of proboscidean species existing in the world, a decline that continues, and now encompasses the remaining members of that family.

One thing that can help us to understand elephants better is having an idea of where they lived and how far they ventured – their range and spread. It is difficult to know the exact prehistoric ranges of elephants, but we do know that the ancestors of both the African (*Loxodonta*) and Asian (*Elephas*) species lived together in Africa for a very long time.

Even after the latter species spread out of Africa into the Middle East and across Asia, some stayed behind until as recently as 11,000 years ago.

Understanding the movement of elephants over the last few thousand years is a challenge, but we can use rock art – like those etchings found in Libya and Sudan – and descriptions from ancient texts to give us some clues. From these historic sources, one thing is readily apparent: until very recently, elephants roamed over most of the African continent, all the way from the Mediterranean coast in the north to the beaches of South Africa.

They spread from the jungles of Senegal in the west, to the highlands of Ethiopia in the east. They inhabited every kind of African environment, from the arid deserts of Namibia and the fringes of the Sahara, all the way up to the treelines of Africa's great peaks. Savannah elephants wandered the slopes of Mount Kilimanjaro and the Rwenzoris, while their forest cousins bathed in the Congo river.

Aristotle spoke of elephants wandering the Atlas Mountains near the Strait of Gibraltar in modern-day Morocco, and Pliny the Elder reported that herds of elephants 'infested' parts of modern-day Libya and Algeria. But by around AD 500, there were no wild elephants living in Mediterranean north Africa.

It has been predicted that up until the sixteenth century – before European exploration, exploitation and colonisation really took hold – there could have been more than twenty million elephants in Africa. This staggering figure is an estimate, based on our understanding of how much space was available for elephants, and how many elephants each of these areas could have supported, given the vegetation they likely contained.

What *is* certain is that from the moment Europeans arrived, the number of elephants in Africa began to shrink. Decline was slow at first, but the causes for that decline 500 years ago are essentially the same as those today – namely hunting for ivory, and habitat loss due to agriculture.

From the sixteenth century onwards, Europeans introduced new crops to Africa, such as maize and sweet potato, which were grown on land newly claimed for agriculture. Elephants would have been using those same areas as natural feeding grounds before they were turned into farms. This increased the number of encounters between humans and elephants, which would have quickly turned ugly if the elephant was intent on sampling the new food source that had suddenly appeared in their stomping ground.

But as well as crops, Europeans also brought guns. And with them, a seemingly insatiable lust for ivory that would change things forever. These European colonisers were not the first to covet and value ivory, though. Long before history was recorded, Stone Age Europeans used mammoth ivory to carve figurines, toys and religious idols. Tomb inscriptions show that Ancient Egyptians had been collecting tusks since at least 2000 BC, and Tutankhamun's tomb of 1325 BC was full of ivory trinkets. The Ancient Greeks also discussed the use and beauty of 'the white gold', long before any of them had seen an elephant's tusk.

The Romans used far more ivory than the Greeks or Egyptians before them, and even in the British Isles, Anglo-Saxon women were buried with ivory accessories. But the procurement of ivory in these ancient times, although a highly valuable trade, was not occurring fast enough to dent the millions-strong elephant population at the time.

Commercial ivory trading really took off in the sixteenth century, at the same time as the spread of agriculture in Africa. This loss of habitat and increase in unnatural deaths had a catastrophic impact on the ranges and behaviour of elephants, and by causing local population declines, made any other losses far harder to recover from.

People didn't only interact with elephants to steal their teeth, of course, and historic accounts show that our fascination with the live animal is not a new phenomenon. The size and strength of elephants has always been a source of wonder for us, and perhaps it is indicative of human nature that it's something we have always tried to exploit.

We know that elephants have been caught, broken and tamed for use from as early as 3000 BC in India and 1500 BC in Syria. Having encountered them in battles in India, Alexander the Great began using Asian elephants around 330 BC like an ancient precursor to the battle tank, and Ptolemy II – the king of Egypt and Carthage – was the first to capture African elephants for use in war. Ptolemy III apparently had an army of around 300 African elephants by 240 BC.

Most famously, Hannibal used a herd of forty African elephants during his invasion of Italy by way of Spain in 218 BC. Three had died before they even reached the Alps, and the remaining thirty-seven were used mainly as pack animals, although they did join in the combat against the Roman cavalry. In the end, all but one died of cold or starvation; and the fate of the last is unclear. Hannibal did use elephants again in a few subsequent battles, until it was made part of his peace terms with the Romans that he surrendered all his elephants and agreed that he wouldn't train any more.

The main advantage of using elephants in war was the sheer intimidation factor – it must have been truly terrifying to see these giant monsters charging towards you across the battlefield at speeds of 25 mph, particularly when they were in large numbers, covered in armour with a turret on top. Even battle-hardened Roman soldiers were known to rout under such an onslaught. However, it's important not to forget that the sights and sounds of battle must have been equally terrifying for the elephants themselves, and if they were wounded or panicked by the enemy, they could end up running amok, trampling men and horses within their own lines.

In fact, because of this risk, elephant riders often carried a large mallet and chisel-shaped tool, which was used to kill the animal if it went berserk and became uncontrollable. Experienced enemy soldiers also learned over time to target the elephants at the outset of the battle – to fire their arrows and direct their lances towards the poor animal's sensitive trunk – in order to make them panic and flee back into their own ranks.

Perhaps because of their unpredictability, Julius Caesar didn't rate the use of elephants in battle, so few African elephants ended up being used in European wars after 47 BC. That said, they are still classed as a pack animal in a US Special Forces field manual issued as recently as 2004! The last supposed use of elephants in war occurred in 1987, when Iraq was alleged to have used some to transport heavy weaponry for use in Kirkuk during the Iran–Iraq war.

Of course, other equally exploitative uses have been found for elephants throughout the ages. Elephants were used as bloody entertainers in gladiatorial tournaments, where they were baited and pitted against other wild beasts. They've even been used as a gruesome method of torture, where the convicted prisoner is either gored or squashed to death by a tethered beast.

There have been many attempts to domesticate African elephants, too. Whilst African elephants should in theory not be any different to Asian elephants in terms of trainability, the culture has not been as deeply ingrained in Africa, so most efforts have generally ended in failure. That said, there was a moderately successful elephant riding school set up in Victoria Falls in Zimbabwe, where tourists could pay to ride, and local mahoots were trained up in the Asian style. However, these trials inevitably met with disaster, after one of the guides was killed in front of horrified guests.

The other major 'use' of elephants has, of course, been as curiosities in zoos and circuses – a trend that sadly continues around the world to this day – and it is estimated that there are up to 20,000 elephants still held in captivity.

A vast range of proboscidean species have walked the earth since their earliest ancestors emerged from the swamps of Africa over 60 million years ago. They have been a remarkably successful group, which dominated the grasslands, forests and tundra of almost every corner of the earth. But only three species survived the dramatic extinctions at the end of the Pleistocene. Humans have been central in the downfall of so many species.

Does the same fate await the last three types of elephants, or can we learn from our history? Before we answer that question, let's have a look at what makes African elephants so special.

3

A Giant's World

There are three species of elephant alive today: the Asian elephant (*Elephas maximus*), the African forest elephant (*Loxodonta cyclotis*), and the African savannah elephant (*Loxodonta africana*), which is the primary subject of this book. Although some debate rumbles on about whether forest and savannah elephants really are two separate species, evidence points to the fact that they are. They look different, their behaviour is somewhat different, and recent research shows that their DNA is distinct, too. These genetics-based studies more or less confirm their separation as two individual species. They likely diverged from a common ancestor as much as 5.5 million years ago, at a time when the climate got cooler and forest habitats became smaller.

The Asian elephant is a more distant relative to the two African species – it's estimated that they split between around 7 million years ago, with the first migration of *Elephas* from Africa into Asia happening some 3 million years ago, although as I mentioned in the last chapter, the two types continued to live alongside each other on the plains of Africa until only a few thousand years ago. Interestingly, genetic analysis has revealed that the closest relative of the Asian elephant is not its African cousin, but in fact the extinct woolly mammoth, from whom they diverged at a similar time to their divergence from the forest and savannah elephant. Visually, these groupings make sense, given the smaller ears, lumpy-looking heads and sloping backs shared by the two species.

Forest elephants, which inhabit the tropical jungles of central Africa, are smaller in height and weight than African savannah elephants, with thinner, straighter tusks that tend to point downwards. They consume much less grass – but more woody plant material and fruit – than savannah elephants, as you'd expect of a species that lives in dense forests, and from what we understand they probably live in smaller social groups, and are much better at climbing steep slopes. The two species overlap only rarely in the fringe areas of Africa's tropical forest belt in places like the Congo.

Asian   African   Forest

Yet across these three surviving species, there are clear similarities that mark them out as elephants: the overall large size, including long limbs and broad, cushioned feet with distinct, flat toenails; the very large head and skull; the trunk; the excess growth of the incisor teeth to form tusks (in many but not all); and the phenomenon of horizontal teeth displacement.

This is a unique process whereby the molars gradually move forward from the back of the mouth to replace worn teeth at the front. There are an amazing six progressions – imagine having six sets of teeth come through! – and the final set of molars, which erupt at about thirty or forty years of age, are enormous, weighing over 3 kg, and measuring 20 cm long and 7 cm wide. The shape of African elephant teeth is how they got their Latin name: *Loxodonta* means sloping teeth.

But the teeth that elephants are most famous for are, of course, their second upper incisors – known universally as tusks. In Asian elephants, the females never have tusks and many males don't either. Some female Asian elephants have what are known as 'tushes'. They resemble very short tusks, but do not have the same tooth pulp inside, and so they never grow further. However, in the two African elephant species, both males and females generally have tusks.

The tusks of males tend to be larger – thicker and longer – than those of females. Mature male elephants in Kenya's Amboseli National Park have an average tusk weight of around 50 kg, whilst the average female tusk of the same age weighs only 7 kg. Tusks grow for the whole of the elephant's lifetime in both sexes, meaning that older elephants tend to have the longest tusks.

Elephants use their tusks in a variety of ways: to help forage for food by breaking or pushing branches and to strip bark from trees. They use them to dig in the ground as they forage for roots or search for water. They are a convenient lever, and are even used to carry things; elephants can often be seen using their tusks to move logs or carry grass in the same way as a forklift truck. I've seen elephants use their tusks as lethal weapons in fights with other elephants and indeed with plenty of other animals that make the mistake of crossing them. With the full strength of a charge, an elephant can easily pick up a fully grown buffalo using its tusks as a spear and whip it into the air.

A fellow officer in my battalion of the Parachute Regiment was once on a military exercise in Kenya. Captain Jay Courtney was leading a patrol through the bush in an area of the country where British troops often went to train before being sent on deployments to Afghanistan or Iraq. It was a warm evening in the autumn of 2018, and Jay led his four-man team on a reconnaissance mission to scout out an imaginary enemy on the far side of a dry riverbed. At around 6 p.m., as the sun was setting over the distant escarpment, the soldiers were in the process of climbing up a sandy ridge onto an open plain dotted with acacia trees and thorny scrub.

Jay was point man, focused on navigating the soldiers towards the position, his mind running through all possible scenarios – would the enemy be laying an ambush, or would they be able to sneak up on them? Did his men have enough water to last in case the patrol was extended? He thought about where he would be sleeping that night, under a thin poncho beneath a star-filled African sky, and wondered whether or not he might get a faint cell phone signal so that he could message home. All sorts of things run through a soldier's mind, especially when he's been immersed in the wilderness for so long, and this exercise had already dragged on for six weeks.

Jay felt comfortable in the bush, and had become accustomed to the ferocious heat, the irritating flies and the daily chore of cleaning his rifle of dust and sand. This was a training exercise, and even though the soldiers carried only blank ammunition, which simply made a loud bang, it had to be treated like a real combat mission and he took his job seriously.

Like all the other paratroopers, Jay was a hardened warrior, used to the rigours of living in the wild. He'd been warned of the dangers of the wildlife and briefed on how to avoid getting on the wrong side of a lion, buffalo or elephant, but so far, he'd only seen the animals at a distance, and generally they ran away from soldiers. Most wildlife has no desire to hang around strange-looking (and smelling) men who like to blow things up and make loud noises. As Jay put it, the elephants were simply 'part of the furniture'.

That is perhaps why, as he strode forward under the weight of his webbing and rucksack, it came as a shock to hear the man behind him start shouting frantically and waving at the confused commander. Jay shook his head and raised an upturned palm to question what on earth the raving soldier was making such a fuss about. You should never raise your voice on a patrol, let alone shout and scream, unless the enemy are already upon you and you're under fire.

Instinctively, Jay ducked and looked around, thinking that perhaps his junior soldier had spotted the enemy troops, or maybe a shot had been fired, and somehow he'd not heard it. But then as he swung back and looked forward, he realised his mistake. It wasn't enemy forces. In this case, it was something far more danger-ous. There, not fifty metres away, was an enormous female elephant with two fat tusks. The giant was shaking her head in anger and flapping her ears, while stomping her feet on the ground.

Jay froze, glancing about him. Now he understood what had happened. Off to the side of the female were more elephants, and babies too. He had inadvertently walked right into the middle of a breeding herd. The mothers, who usually corral their youngsters into the middle of the herd to protect them from predators, had not spotted Jay. The wind had been blow-ing towards the patrol, so the elephants hadn't picked up the soldiers' scent before it was too late. Jay was almost surrounded, while the other three men scarpered to the safety of the edge of the riverbed.

For Jay, the moment seemed like an eternity, but it couldn't have been more than a few seconds. The matriarch, startled by the presence of an armed human, was not about to take any chances with so many vulnerable calves in the herd. She gave him one ear-splitting trumpeting call as a final warning before charging at full speed right towards him. As the elephant crashed through the bushes at twenty miles an hour, Jay's mind spun in disbelief at the surreal vision that unfolded in front of him. This doesn't happen except in the movies, he thought, surely it will stop soon?

It was too late to run anywhere and there was nowhere to hide. In any case, he remembered what the brief had said, you can't run from an elephant. His attention became fixated not on the bulk of mass hurtling towards him, but instead at the glistening ivory tusks that appeared like spears before him. They were cracked and patchy; one was slightly longer than the other, and, he thought to himself, decidedly blunt. Even so, the primal fear inside him took over as his body exploded with adrenaline, and he knew that whatever happened, he must avoid those tusks.

The rest is a blur, but from the accounts of the other soldiers who watched on in horror, the elephant smashed into Jay with her forehead, sending the young officer flying into the air, before the 3-tonne beast continued with the assault, ramming her tusks into the ground either side of the man. Jay remembers seeing the enormous grey hulk hovering above his head and stamping down on the earth, all the while using her trunk to flick and toss his injured body around like a piece of cloth.

He tried to crawl away from the carnage, but there was no stopping it. Then the elephant delivered her message home in the only way she knew how, by driving one of her three-foot-long tusks right through his arm, ripping the paratrooper's limb almost in two. Despite its rounded end, the tusk speared his tricep, tearing through the flesh and muscle like a hot knife through butter. Jay's body crumpled, and the only thought he could muster was that it might be better to play dead. He curled into a ball and hoped above all else that he might survive.

The elephant calmed down, her bloodlust perhaps satisfied by the thought that he no longer posed a threat, but she still loomed over him, her weight crushing his chest. Then, as the creature was about to stamp on his lifeless form, the other soldiers began to fire their weapons in its direction. Despite the fact there were no actual bullets flying, the noise seemed to do the trick. The elephant, startled by the bangs, raised her trunk and trumpeted again, reversing backwards and flicking her head in disgust. The men jolted forwards, shouting and firing more shots into the air, until the creature backed off and slinked away in the direction of the rest of the herd. A few seconds later they had disappeared, and all was silent.

Luckily for Jay, he survived. The other soldiers ran to his side and held his smashed arm together, applying a tourniquet and stemming the flow of blood until the medics arrived at the scene ten minutes later. He was rushed to hospital in Nairobi, where doctors managed to put his mangled arm back together, and now he's left with only an impressive scar and a good story to tell. One thing's for certain, though, he won't ever think of elephants as merely part of the furniture again. And it's a lesson that anyone who walks on foot through Africa is advised to keep well in mind: those tusks are not just for show.

Apart from their tusks, the other defining feature of elephants is obviously their trunks. Formed by the fusion of the nose and upper lip, trunks are truly amazing appendages. Imagine having a six-foot-long nose that doubles up as a hand.

Trunks are used for gathering and picking up food and other things with remarkable precision. African elephants have two pincer-like 'fingers' at its tip, one at the top and bottom, whereas Asian elephants have only the one at the top. Trunks can be used to push over and break up large food items, sometimes as big as whole trees. They're also handy for trumpeting, dusting, scratching, bathing and snorkelling, as elephants are born swimmers. It's an enchanting sight to watch young elephants tumbling around in a watering hole, as they learn how to spray water and play with each other.

As Jay found out, the trunk is also a formidable weapon with which to smash others, and even to launch projectiles: elephants can throw things like sticks and stones, and with pretty good aim. And, of course, elephants use their trunks to drink with, sucking up litres of water before pouring it back into their mouths.

Trunks are used almost continuously to check on the rest of the social group, either in a tactile way – reaching out and touching others by way of greeting or reassurance – or by sniffing others to get information. Because above all else, trunks are fundamentally a nose – a highly mobile and phenomenally sensitive nose, at that.

The mobility and precision of the boneless trunk comes from the 40,000 or so muscles it contains. By way of a comparison, the whole human body contains just 639. The muscles of the trunk are divided into more than 100,000 fibre bundles, each served by a mass of nerves and connective tissue. And to illustrate how sensitive a nose it is, consider that elephants have five times more olfactory receptor genes than humans and more than twice as many as dogs. These are the genes associated with our sense of smell and while this does not necessarily mean that an elephant is twice as good at smelling stuff as dogs are, it certainly means they have more sensitivity to a broader range of scents.

In fact, recent experiments with Asian elephants have shown they can *smell* the difference between buckets containing either one or three scoops of sunflower seeds, correctly choosing the bucket with more food. Clearly, we could do the same by looking, or feeling the weight of the buckets. But by smelling only, through a sealed lid? I don't think we'd stand a chance.

The importance of scent and smelling to elephants becomes very apparent when we begin to look at their brains. Elephants dedicate a huge area of their large brains to perceiving and processing smells. The size of various brain parts shows us that hearing sounds and producing vocalisations is also of considerable importance to elephants, whereas vision seems to be much less significant, with the areas of the brain that process visual signals being much smaller than those that deal with smell and sound.

This confirms an important point about elephants: they must perceive the world in a rather different way to us. As a species, we rely heavily on sight to get information about the world around us, but for elephants, smell and hearing are the dominant senses.

When you scale the brain against body size, humans win out. We have the largest brains relative to body size with bottlenose dolphins and chimpanzees coming after. An average person is seventy-five times smaller than an adult male savannah elephant, yet our brain is only three or four times smaller than that of an elephant. That said, when it comes to absolute terms, there is no brain bigger on land than that of the elephant. They weigh in at up to 5.5 kg in males and slightly less in females. Sperm whales and orcas do have larger brains than this – at around 7 kg – because water can support a heavier head and body. Humans by contrast have a brain that weighs just 1.3 kg.

Compared to many other mammals, the brain of the elephant is located low down in the head – in line with the eyes – while the upper skull is formed of a honeycomb-like bone structure with air pockets that reduce the weight of the head, but keep structural integrity in place. This adaptation has been the fatal undoing of many a novice hunter, who has aimed too high when attempting to kill an elephant, succeeding only in wounding and enraging the poor animal.

Aside from the enlarged olfactory and auditory regions, the elephant brain also has an especially well-developed cerebellum. All mammals have a cerebellum, which is mostly involved in overseeing and coordinating movements and voluntary muscular activity. But in elephants it is huge, with many more neurons, organised in a much more complex way, compared to other mammals – including us. This makes sense, given the complexity and range of movements that the trunk is capable of performing. So elephants have brains that are specialised for smelling and hearing, with very fine motor control.

Most mammals are born with brains that already weigh around 90 per cent of what they will do when fully grown. This means that they are almost fully developed. The brain of a newborn human, in contrast, weighs only 25 per cent of what it will do as an adult.

If you think of the 'childhood' of the average mouse or horse, versus our own species, this makes sense. Those species are up and running soon after birth, whereas humans are pretty useless for the first year or two after birth. But a newborn elephant brain weighs 35 per cent of the adult brain weight – the same percentage as chimpanzee infant brains, and slightly lower than the 40 per cent figure for bottlenose dolphins. The brains of all these species have a lot of developing to do during their long childhood – almost as much as we do – suggesting they also have a lot to learn.

An elephant is big. Very big. A fully grown adult male can reach a shoulder height of 3.4 metres and weigh up to 7,000 kg. That's the same as four family cars, with passengers. The largest known elephant was a male, shot in Angola in 1956, which was a colossal four metres tall, as big as some of the prehistoric species we mentioned, and is thought to have weighed 10 tonnes. And while there are other megaherbivores roaming the earth today – including four species of rhinoceroses, the common hippopotamus, and giraffes – none of them come close to the size of a full-grown African elephant.

The elephant's enormous size has been the key to their evolutionary success. It gives them a triple whammy of survival benefits: making them less vulnerable to predators; enabling them to live in a wide range of habitats; and meaning they can eat a wide range of foods. This allows elephants to move into different areas when there's not much food around.

All the same, there's a big difference in size between male and female elephants, with females normally around one metre shorter than males, and topping out at around 3,000 kg.

The difference in size between males and females all comes down to sex. It's a common feature in many animal species, and is known as *sexual dimorphism*. In elephants, as with many mammals, the males don't have much to do with raising their offspring. Their role ends basically at finding and then competing for receptive females with whom to mate. Male elephants aren't particularly choosy when thinking about the size of their partners, either, and as long as the female is in *oestrus* (the time when they can get pregnant and so are 'sexually receptive'), males will try to mate with them. That means there's no selective pressure picking out larger females over time. The only thing limiting the number of offspring a male elephant can sire is the number of females he can find and mate with, which in theory could be in the hundreds.

For females on the other hand, as is often the case, size does matter. Females put a lot of time and energy into raising their young, so they want to give each one the best chance of survival. This means they want the healthiest and strongest males to breed with, in order to ensure their offspring are more likely to be healthy.

It takes twenty-two months of pregnancy for a female elephant to produce a newborn, which can weigh more than 100 kg at birth. The female spends the next few years nursing the youngster (known as a calf), until it can be persuaded to give up milk and survive solely on solid food – often only because the mother has had a new calf that needs her milk more. So, there is usually a gap of around four or five years between each new baby. This means a female elephant who typically gives birth to her first calf when she is between the ages of eleven to fourteen years, and who lives to sixty, might raise only nine or so calves over her lifetime.

Sexual Dimorphism in Elephants

Based on the numbers alone, a female has fewer chances of offspring successfully passing her genes to the next generation, so she invests heavily in those chances. Males, on the other hand, play the numbers game – going for quantity over quality. This in turn sets up competition between males for the attentions of likely females. It usually goes no further than intimidation, because physical combat runs a high risk of injury or even death. But, if neither one backs down, male elephants will often fight for the right to mate with a female.

Ultimately, that means the largest, most dominant bulls are more likely to mate successfully than smaller males. Their genes are passed on to the next generation, and the process of natural selection therefore favours larger males. Such sexual dimorphism is common in species that compete in this way for access to mates, but elephants have one of the most pronounced size differences amongst any mammals.*

Whatever the size difference between the sexes, elephants are still gigantic when compared to other land animals. This has influenced everything from their diet and their habitat, to their effect on the environment, as we shall discover next.

---

* The champion of the mammal world when it comes to disparities in body size between the sexes is the southern elephant seal – where mature males can weigh up to 4,000 kg, which is seven times greater than an average adult female elephant seal.

4

The Ultimate Survivor

It's amazing how an animal the size of an elephant can simply disappear. Once, I was in Uganda with my guide Boston, and we'd been tracking a breeding herd of twenty or so elephants along the Nile river. They were standing in a clearing about ten yards away, grazing on acacia bushes and stripping the branches bare of their leaves.

The matriarch knew we were there, as she kept raising her trunk in our direction, sniffing the wind. She was calm, though, so long as we kept our distance and tried to stay downwind of the herd. With her was a newborn calf. As the soldiers in Kenya discovered, this can make possibly the most dangerous combination, as a mother will destroy anything, or anyone, that comes near her baby. But in this instance, I felt safe. There was enough cover and Boston reassured me that we were fine, so I followed him as we crept along a trail.

'Look over there,' he pointed.

I stared into the thick bush ahead. 'What?'

He smiled menacingly. 'Those bushes. In my language we call it the devil's armpit.'

'Why?'

He chuckled, 'Because it's scraggy and bushy and can hide all sorts of nastiness.'

It was remarkable to see these enormous animals, whose backs reached the top of the bushes, stripping all the leaves clean off them.

Then suddenly there was a noise in the distance behind me. It was the faint call of a lion grumbling a mile away. I turned around and tried to listen for the king of the jungle, but after ten seconds there was nothing, so I turned back around.

Looking up, I stared back at the clearing. Where there had been a whole herd of elephants, now there were none. No noise, no movement, nothing. It was as if they had never existed.

I was astounded, 'How on earth did that happen?' In my naivety I suggested that we might follow them out of curiosity.

Boston laughed. 'No, they are gone, they will hide now, and anyway, we never go into the devil's armpit, because even if you walked a metre away from one, they would be invisible; then one false move, and you'd be tusked and stamped on. An elephant is 7 tonnes of silence. You'd be food for the lions before anyone even found your trail.'

Elephants are huge, which makes their silence and ability to blend in all the more surprising. Their size has affected every part of their biology, from how much they need to eat and drink and how they keep their temperature regulated, to where they live, what they feed on, and how successful they are at producing calves. It's all about eating, heating and sex.

It seems obvious that the bigger an animal is, the more energy it needs over the same period of time. This is what we call the *metabolic rate*. But the metabolic rate doesn't increase at the same rate as size does: a 6-tonne elephant needs only 400 times the energy of a 2-kilogram rabbit, not, as you might speculate, 3,000 times the energy. Bigger animals are more efficient with their energy than smaller ones, particularly at conserving heat. That's because, as an animal gets bigger, its volume grows at a faster rate than the surface area of its skin.

The stomach and intestines of a bigger animal also take up a proportionately larger amount of its body. That means food spends more time in their gut, so they are able to extract more energy and nutrition than a smaller animal and still get all the nourishment they need. The upshot is that large animals can eat lower-quality forage than a smaller animal. An elephant needs to eat only 5 per cent of its body weight every twenty-four hours, whereas the smallest antelope species in a similar environment – the dik-dik – needs to eat 20 per cent of its body weight.

They might not be precious eaters, but to meet their enormous nutritional and energetic demands, elephants need to eat a lot, and can be found feeding for around eighteen hours a day given the chance. They eat almost constantly with only short periods of rest, usually in the early hours of the morning and around midday when temperatures are at their highest.

Elephants have evolved into the ultimate herbivore generalists. Whereas most herbivores tend to be either grazers (who eat grass) or browsers (who eat woody plant material), elephants scoff the lot. They can eat everything from fruit, leaves, fresh shoots and flowers, to grass, roots, branches and bark. What they eat depends on what is around, and they switch between browsing and grazing depending on the time of year.

The savannah of sub-Saharan Africa broadly has two seasons: the cooler, dryer winter; and the warmer, wetter summer. The first rains at the beginning of summer lead to a flush of green grass, creating an abundance of food after the comparatively lean, dry winter. During summer, elephants get more than 90 per cent of their dietary requirements from grass, but during winter, when grass availability and quality drop dramatically, they switch to browsing, and target a wide range of woody plants from large savannah trees to small shrubs. They eat bark, dig up roots and tubers, and I've been told they have even been seen to munch on the grassy nests made by weaverbirds.

While the bulk of an elephant's food comes from a few abundant favourites, some elephants have been known to eat more than a hundred different plant species. Being unfussy eaters stands elephants in good stead to find sufficient food to keep themselves going across a diverse range of habitats and in all seasons.

Both males and females need to eat a large amount, but the way they meet their daily calorie requirements is quite different.

Male elephants become independent and move away from their family groups as they hit adolescence, usually around fourteen years old. By this age, males also start to grow much more rapidly than females, who tend to reach their maximum height by the age of twenty, whereas males keep growing throughout their life.

Remember, this larger size means that males have a lower metabolic rate, so need less energy per kilogram of body weight, and can eat even lower-quality, more fibrous food than female elephants. Males feed on taller trees than females, eat parts of the tree that females would turn up their noses at, so to speak, and are much more likely to push a tree over, strip away the bark and gorge on whatever is left.

Not only does their size make the males more powerful and destructive, like in many mammal species (including humans, allegedly), male elephants are more likely to engage in risky and adventurous behaviour. These two make a fiery combination, sometimes witnessed when they come into contact with the human world. It's usually the males that go off raiding in farmers' fields, stealing the highly nutritious and no doubt tasty crops planted by people.

Crop raiding allows a male elephant to get a maximum nutritional return for minimum time and energy – possibly getting up to almost half of his nutritional and energetic needs in only 10 per cent of the usual foraging time; a clever move, sometimes worth the risk, especially if he needs that energy to find a mate and reproduce. Of course, the strategy comes with a significant danger of injury or death, as farmers and wildlife managers designate them as rogue and often kill these 'pest' animals. But more on that later.

Females, however, tend to be more cautious and stick with their family group. They're more sociable and spend time interacting with each other. As a result, their mealtimes are frequently interrupted, given that there are more individuals around to create distractions. Motherhood also places extra energy demands on females. For the first nine months of its life, a calf will consume up to 13 litres of milk a day, and a mother's energy requirements go up drastically. So, females have to find higher-quality foods than males to counteract these issues, and to provide suitable food for calves that are beginning to add solids to their diet.

Consideration of calves is so ingrained in adult females, they are even known to reach for higher food when they are near youngsters. This extra effort means that enough food is benevolently left lower down in the canopy for the juveniles to eat. Calves are small-bodied and growing fast, so they need all the help they can in finding the best quality food. As calves, male and female elephants forage in the same way. It's only in adolescence, when the males quickly get bigger, and the females prepare for the demands of reproduction, that their feeding behaviour starts to diverge. But while they are young, it makes them highly vulnerable during periods of scarcity such as a drought, and the mortality of young elephants is one of the main natural controls of elephant numbers.

The size of area that an elephant uses to meet its nutritional needs and allows successful reproduction is known as its 'home range'. Its extent depends entirely on the environment, and the variations can be enormous. In areas with high rainfall and plenty of resources – such as Lake Manyara National Park in Tanzania – their annual ranges can be less than 100 square kilometres. Contrast that with the arid desert environment of Namibia, where home ranges extend to 14,000 square kilometres. However, the all-time record belongs to a female elephant from the Gourma population in Mali that walked across 32,000 square kilometres in a year.

As food becomes scarcer and watering holes dry up with the changing seasons, an elephant's range changes. They often have larger ranges during the wet summer than in the dry winter, when they have less energy to travel far. Both males and females are likely to be sexually receptive during the wet summer, because there is more food around to support the energetic business of finding mates, going into oestrus and, of course, growing a baby.

Because they are socially independent, particularly when on the hunt for a female, mature adult males have much larger roaming areas than breeding herds and they move much faster. Males commonly go into a heightened sexual energy known as *musth* in the summer. It's a period marked by surging testosterone levels that can last for several months, swollen and weeping temporal glands on the side of their head, and the dribbling of urine that has a very strong, distinct odour.

During musth, male elephants can often become aggressive and highly frustrated; their sexual energy overflows and they become desperate to find receptive females. Musth males have one thing on their mind, and they don't mind who or what they have to fight to get it. Or indeed how far they have to walk to find it – males in Amboseli, Kenya are often recorded walking over thirty kilometres a day repeatedly to find females. And it's the older males that tend to be the most vigorous, travelling fast over very large areas. All this marching around requires a lot of energy (hence why it's typically done in the food-rich summer), but it significantly improves a male's chances of passing on his genes.

In the dry winter season, adult males will shrink down their range and move around far less. They tend to focus on patches of woody vegetation that provide abundant, but low-quality food, and rely on their considerable fat reserves to see them through the leanest time of the year. Females and their family groups have less flexibility, because of the constraints of social living and their need for higher-quality food. So, their ranges have to continue to include a greater variety of feeding areas.

In order to find all this food, elephants need to be excellent navigators, and their spatial memory is rightly celebrated. Food and water sources can be distributed patchily over a huge area, and they might be useful only at a specific time of year. It's very impressive that elephants can go months, or even years, between visits to a particular location, and need to remember *when* a food or water source is worth visiting, not simply where it is.

A GPS tracking study in the arid scrub of Etosha National Park in Namibia showed that elephants made very direct and intentional movements towards a specific water hole from up to fifty kilometres away. Not only were the elephants able to navigate directly to these specific locations, but they invariably chose the water hole that was nearest to them, demonstrating the kind of detailed navigation that these days I can only achieve with the aid of electronics.

Finding water is a vital consideration in an elephant's wanderings and their distribution across the savannah is mostly determined by where it can be found. Fully grown adult males need well over a hundred litres a day, so regular access to water is crucial for an elephant's survival. They commonly drink at least once a day during the wet summer, and every two or three days during the cool season, when many watering holes dry up. Smaller-bodied calves who tire easily, and mothers nursing their young with milk, both need lots of water every day, so family groups have to stay relatively close to watering holes.

This requirement of family groups to stay close to water can have disastrous consequences. In the early 1970s, Tsavo National Park in Kenya suffered a serious drought. Surface water remained in some areas, like the Galana River system, but two years of low rainfall had seriously reduced the amount of grass and edible tissue on woody plants. The meagre amount of forage near the remaining surface water was quickly eaten and many female elephants, trapped by their instinct to stay close to the water, starved to death. It wasn't any better for the males either. Those who dared to walk further away to find food became so dehydrated that when they returned to the river and drank desperately, their stressed bodies ended up shutting down completely with this sudden, large intake of water and many perished. By the time the drought lifted, over 6,000 elephants had died.

This may be merely a taste of what is to come. Climate change will have a massive impact on sub-Saharan Africa, with more frequent extreme heatwaves and a major change in rainfall patterns, which will lead to a drier climate in southern Africa and a wetter environment in tropical East Africa. These changes are already being seen, and will have profound implications for wildlife. The available range for elephants will inevitably shrink over the coming decades, as habitats on the edge of the range become uninhabitable. Sadly, as these roaming zones shrink, more and more elephants will die in tragedies like the Tsavo drought.

Aside from feeding, control of its body temperature (known as 'thermoregulation') is the other most important factor in determining what an elephant does with its time. The average body temperature of an elephant is very similar to ours, at 36.5°C, and it's critical for an elephant to maintain that within a fine margin to prevent overheating or getting too cold.

An elephant has less skin surface area proportionately than a smaller herbivore, so they are better at conserving heat. However, in summer the savannah regularly sees temperatures soar over 35°C, making it more important to *lose* excess heat. This is compounded by those large exposed bodies, which absorb heat directly from the sun and surrounding landscape. To counteract this, elephants have evolved behavioural and physiological tactics to keep cool.

Their main thermoregulation strategy is to get out of the sun during the hottest part of the day. That's why you'll see family groups and independent males looking for shaded woodland areas as midday approaches. Under the shade of trees, they can rest and keep cool. If they can find water to bathe in, that cools them down even faster.

Elephants are accomplished swimmers, despite their considerable size and weight, often using their trunks as a natural snorkel. As well as using rivers and lakes to cool off, elephants can easily swim across open expanses of water to escape potential threats, or to reach a safe spot where there is access to food. Even better than swimming, mud wallowing is a really efficient way of cooling down, and elephants in the salt pans of Botswana have been spotted carrying water from one place to another to make sure they can always wallow in their favourite mud holes.

Elephants don't have sweat glands, but they do lose water through their skin from evaporation. This can account for up to three-quarters of their heat loss, but that water needs to be replaced, hence their need to drink so much, so often. Then, of course, they have those ridiculously big ears – particularly on African elephants – that make up around 20 per cent of their total skin surface area, and play a very important role in preventing overheating. The ears are laced with blood vessels that can carry up to 18 litres of blood a minute, and their size, manoeuvrability, and comparatively thin skin make them a very effective heat-exchanging device. They are, effectively, giant air conditioners.

The flow of blood to the ear can be controlled by the dilation and constriction of the blood vessels, depending on the weather outside. During the hottest part of the day, blood is pumped into the ears. The liquid going in is three degrees warmer than the blood coming back out of the ears, showing how effective they are as a cooling system. When the air temperature is really hot, elephants flap their ears back and forth to make this work even more efficiently. When the weather is cooler or it's raining, elephants hold their ears flat against the head and constrict the network of blood vessels. This reduces inward blood flow and limits heat loss from the body.

The large ears of the elephant are a remarkable adaptation. They allow elephants to tolerate temperatures in excess of 40°C and enable them to remain active during the hottest part of the year – the peak of the wet season. This is vital if elephants are to get enough food to meet their energy demands. The role of the ears in thermoregulation also gives us a possible explanation for why mammoths, who lived in the cold tundra landscapes of the northern hemisphere, had much smaller ears than their cousins in the savannah. Mammoths had the opposite challenge of maintaining their body temperature by conserving heat at all costs.

Some ongoing research suggests that African elephants may also even use their erect penis – which can be almost two metres long – in the same way as their ears, to cool down on very hot days by getting it out, so to speak, and splashing it with water. Perhaps mammoths had the opposite problem in that department, too.

Given their food, water, and temperature requirements, it is a wonder that elephants can survive in deserts, and yet some do, living right at the limit of what the species can endure. The desert elephants of Mali and Namibia regularly go for days without water, while experiencing daily temperatures that frequently top 40°C. These elephants have evolved some unique behavioural adaptations to survive in such an extreme environment. Within their vast ranges, they seek out small patches of food using their excellent memory. In times of extreme drought and food scarcity, young desert elephants even ingest the dung of adults* to get supplementary nutrition and water. Elephants there have been spotted urinating on mud and splashing this over themselves when wallowing holes are dry, and they can also cool themselves by spraying water stored in the pharyngeal pouch when surface water is not available.†

As well as influencing their daily rhythms, the sheer size of elephants and the amount that they need to eat has a profound impact on the landscapes they live in. Elephants affect their neighbourhood so much that they are classified as a *keystone species*. That is, one that has a disproportionate effect on the ecosystem it inhabits. Other examples of keystone species are the wolves of Yellowstone National Park, lions, and sea stars.

As elephants forage, they consume vast amounts of plant material, uproot entire trees and dominate water sources. This incredible ability to change their environment completely and to have an impact on the availability of resources is greater than any other vertebrate, except humans. It's an impressive sight to watch an elephant uproot a tree, wrapping its trunk around the tree to bend and weaken it, and pushing with its mighty strength – they make it look so effortless.

In areas that have high densities of elephants, their intensive foraging can even lead to the disappearance of their favourite

---

* A practice called coprophagy.

† The pharyngeal pouch is a pocket of tissue located at the top of the digestive tract, and this behaviour has only been seen in other elephants as an emergency cooling technique after a stressful encounter that might cause them to overheat, such as fleeing from danger. Desert elephants are the only ones that use the pharyngeal pouch to cope with ambient high temperatures and a lack of available surface water.

tree species, and as elephants have been confined to ever smaller areas, there has been concern that in some parts of Africa they are damaging and degrading the savannah to an irreversible extent and in doing so, reducing biodiversity.

But elephants do more than simply destroy. The African savannah is a complex ecosystem, defined by two plant forms that are competing for resources: grass and trees. In most environments, only one of these forms dominates: grass in the case of steppe, and trees when it comes to forests and jungles. In the savannah, they coexist in an uneasy balance, with the winner at any one time determined by rainfall, soil nutrients, fire, and which animals are eating them.

Whilst uprooting and eating trees seems destructive, it helps other species in many ways by creating more complex ecosystems: in northern Kenya and north-east Tanzania, frogs and lizards are more abundant in areas that have more elephants feeding. This activity also makes it easier for smaller herbivores to get at food that would otherwise be out of their reach. And the sheer volume that elephants eat helps, too, because it encourages new plant growth, of a higher quality.

By opening up dense thickets, elephants even help lions and leopards, as it allows them to get into areas favoured by small herbivores. But it doesn't stop there: by digging with their tusks and feet, elephants enlarge existing water holes, and excavate new wells when they access sub-surface water and mineral-rich sediments. The movement of elephants across the savannah has also created defined 'highways', which are subsequently used by all sorts of other creatures for many years to come.

Even an elephant's dung has an impact on the environment. Because their diet is so fibrous, only 50 per cent of the plant material will be digested. Given how long food stays in an elephant's digestive system, and the ground they can cover in this time, their dung can distribute huge quantities of nutrients and seeds all over the savannah. Digestion helps the seeds that elephants swallow to germinate, whilst the nutrients in the dung provide the seeds with everything they need to get established. So elephants are deemed to be central in maintaining plant diversity and distribution, giving them the nickname, 'mega-gardeners'.

Whilst elephants do damage areas in which they are over-crowded (often due to fencing-in or human-driven habitat loss), when left to their own natural devices, they cleverly maintain the health and diversity of dynamic savannah, grassland, and forest habitats. Their forestry techniques are part of a finely balanced ecosystem and anything that happens to them has a knock-on effect. Without this keystone species, the entire ecosystem could disappear. The behaviour of the elephant is hugely complex, and we are only starting to understand their pivotal role in the environment and what that means for the other species with whom they share the landscape.

It's becoming very clear that these giants have a fundamental part to play in nature and much of this is down to their enormous size. But their importance doesn't stop there. Elephants aren't simply big eating and breeding machines, they also happen to be highly intelligent, sociable creatures that rely on love, relationships, and interaction far more than we might think.

5

Friends and Relations

Sometimes elephants go wandering off alone or in small groups away from the herd. They may have found a particularly tasty treat they want to eat, or they might have had a fight with a relative and sulked off with their tail between their legs for a while. It's common for elephants to do their own thing for a night or two, but more often than not they go back to find their family or friends, and one of the most heartwarming sights I've seen in Africa is when elephants are reunited after time apart.

Once, when I was on a horseback safari in the Kenyan highlands, I spotted a herd of elephants grazing on the side of a hill on the far side of a river. As we were cantering along, I noticed three more elephants suddenly emerging from a eucalyptus forest to my right. At first, I didn't know if they were two distinct groups; perhaps they were competing for food, and I wondered if there would be trouble. To start with, the two parties raised their trunks, sniffing the air to confirm whether they were friends or not, and I pulled on the reins to stop and observe.

The newcomers dramatically rushed forwards, trumpeting, flapping their ears and shaking their heads, as they splashed right through the shallow river and ran up the hill towards the other group. I prepared myself for the commotion and feared for the safety of a tiny calf, who began to whirl around chaotically. I needn't have worried. As it turned out, they seemed to be on rather good terms. The matriarch of the herd, the oldest and most experienced female, led what can only be described as joyful celebrations, signalling her welcome with a series of loud calls, as the rest of the group followed in the excitement.

Some of the younger calves began spinning around and running in circles and I noticed that many of them were even peeing themselves in a state of raw excitement. When the two groups got close enough, they all began touching and caressing each other with their trunks, in almost the same way that people shake hands or kiss. Clearly, elephants have no problem showing a bit of public affection! The outpouring of emotion was quite unashamed and I felt no doubt in my mind at that moment that these magnificent beasts were happy, in the purest sense of the word, and must have missed each other when they were apart.

Relationships are what make elephants tick. Just as their large body size is closely linked with how they feed and move around, their need for interaction and companionship is fundamental to understanding the nature of an elephant and how it behaves.

Feeding and ecology – what they eat, where they move, when they rest – is clearly very important to appreciating elephant behaviour, but it would be impossible to comprehend elephants properly without also thinking about their social lives. Elephants may spend three-quarters of their day eating, but with the exception of mature adult males, they are hardly ever doing so alone, instead being almost constantly in the company of relatives or friends.

These relationships are so strong that elephants have frequently been reported as neglecting their own nutritional requirements for many hours or days to remain close to ill or dying companions, often risking severe dehydration to keep circling scavengers at bay. Their dedication to each other is truly remarkable.

Most mammals are not social – at least not in the sense of permanently living and socialising with the same individuals, day after day and year after year. But elephants are not most mammals. Elephants have a very strong sense of society, the foundation of which is the mother–infant bond. All relationships, associations and behaviour radiate out from this core, even though as adults, males and females have rather different social lives. For females, life continues to centre on the family; whereas for males the dynamics may change with age and sexual status, but their relationships with others remain crucial.

Female calves typically remain with their mothers for life, resulting in matrilineal family groups composed of mothers, sisters, aunts, daughters, granddaughters, nieces and cousins of multiple generations. These close-knit families are characterised by persistent and strong social bonds between individuals, and these individuals typically coordinate their activities so that they are all feeding, moving, resting or socialising together.

Families within the Amboseli National Park population of Kenya, undoubtedly the best-studied elephant population in the world, have been recorded as containing anything up to fifty elephants, although they typically number around twenty individuals with an average of seven adult females.* Whilst these families are very close – both socially and genetically – membership is not always static.

* Interestingly, Asian elephant social organisation is somewhat looser, with smaller and less coherent core groups and less connectivity at the population level than African savannah elephants.

Female elephants occasionally choose their own families, and there are some unusual cases of unrelated females being fully integrated into a new herd, and other odd examples of females splitting off with unrelated females to form new groups, but it's very rare indeed to see a female elephant alone. They are highly sociable creatures and seem to need the company of others to thrive.

Elephant families grow or shrink in number because of births, deaths and the departure of adolescent males, but when individual females do sometimes split off and join with others it is usually only temporary, and with good reason. Splits allow a family to cope with varying environmental conditions such as seasonal food shortages, because they can separate to search for food or water in different areas, and then reunite later when they have had enough to eat, or have found a patch of food that can support the whole group. Interestingly, this type of social organisation is also shared by some whale and dolphin species, chimpanzees, several monkeys, and of course, human societies.

What's more, individuals from one family may associate with elephants from other families, often blood relations, and together these two or three related families form a bond group. It doesn't stop there, though. The layers of female society build up further with several bond groups forming a clan, and multiple clans making a sub-population, which in turn create regional populations. Some groupings meet up frequently, and others only occasionally, all subject to the size and density of the population and the availability of resources.

Adult females are constantly thinking about where to eat and drink, how far to walk and who they want to spend time with. That's a lot of decision-making, and as in any society, too many cooks spoil the broth. If all the adult elephants in a breeding herd had their own way, then chaos would doubtless ensue as the females argued and acted in their own interests, or in the interests of their own calves. Instead, somehow they manage to cooperate, act in unison, and in a very efficient manner. So how is this achieved?

Elephant society is generally viewed as accommodating, inclusive and egalitarian, but what holds it all together is solid leadership, and when there's a need for leadership, it usually comes from the matriarch. For most actions, most of the time, it's her decision that's final. That's not to say there isn't a fair amount of negotiation, and when other females can't abide the boss, they can always storm off temporarily, or even for good. But for the most part, individuals in breeding herds respect their leaders, and will follow them anywhere.

As I walked with herds across Botswana, it was fascinating to see the courage and leadership demonstrated by matriarchs in all sorts of different scenarios. In the dry season, elephants are forced to travel vast distances across barren salt plains and parts of the Kalahari Desert. Driven on by thirst, they march hundreds of miles across the parched plains. The matriarchs lead the way, having made this journey countless times before. But these elephants are not the only animals in search of water, so on find-ing a watering hole, they often have no choice but to share it with all kinds of other wildlife, even lions, who are not averse to preying on young elephant calves.

I remember watching with the utmost respect as a pride of lions approached a pool on the edge of the Makgadikgadi, where one herd of elephants was drinking at dusk. By day, the elephants dominate the water hole, but as night falls, the power balance shifts; with their poor eyesight, the hulking creatures are at a disadvantage, and the lions use the cover of darkness to sneak up on anyone wandering too far from the herd.

As soon as the matriarch caught the scent of the lions, she immediately rallied her family, resorting to pushing and shoving the babies into the middle of the herd and bossing the other females into action to form a huddle, protecting the little ones from the danger. Only when a solid defence was formed did the herd move off, the matriarch always keeping herself between her kin and the predators. It was this kind of defensive behaviour that the soldiers in Kenya had encountered, and another reminder of why one should never get in the way of an elephant mother and her young.

Female elephants can spend forty or even fifty years producing and raising calves. Yes, that is *five decades* rearing children. As mentioned earlier, the female will also be suckling her current calf throughout the four- or five-year period between births, with the older one often being stopped from nursing only when the newborn arrives.

These figures and time periods all add up to one thing: a female elephant can spend virtually all of her adult life either pregnant, lactating, or both.

Even with this colossal dedication to raising young by any one female, the old adage of 'it takes a village' is perhaps truest for elephants. The whole family shoulders a huge amount of responsibility for all calves born into it. Juvenile females will usually babysit calves, following them around, touching and caressing them, steering them away from danger and keeping them close to the family group. Younger, less-experienced mothers seem to follow and learn from the older, wiser females how best to raise their babies.

Motherhood is a steep learning curve for elephants, even those with a lot of babysitting experience. As we might expect, whether a young calf lives or dies depends to a large extent on external factors such as droughts, predator attacks, and conflict with humans. But it also depends very much on the knowledge, experience and social situation of the mother.

A mother has to understand and assess the needs of her calf. She has to know when the demands for milk are necessary, or merely greedy. She has to know when the 'distress bellow' is genuine or only playful; or when the calf really can't keep up with the pace anymore, compared to when it is being a bit lazy. First-time elephant mothers are often not very good at it.

A calf from a large family with a lot of young females that can act as babysitters, as well as older females that can provide guidance, is much more likely to survive past the age of two than a calf born into a small family without many female relatives. In fact, adult females with surviving mothers are more successful in their own reproductive lives, producing calves sooner and for longer than those whose mothers are dead.

Having a grandmother on hand also makes elephant mothers better. And of course, the age and experience of the best grandmother of all, the matriarch, is hugely important. Calves born into families with older matriarchs have even better chances of survival. Calves – and their mothers – need sisters, cousins, aunts and experienced grandmothers to help them make it to adulthood.

Even with all this help from family members, calves remain dedicated to their own mothers, and up to around ten years of age will frequently call to her and touch her, reaffirming their bond. For the first couple of months of life, infant calves rarely tend to be more than a trunk-length away from their mother. This distance gradually increases as they grow in size and confidence, but there are some differences between male and female calves.

Females move away from their mothers towards other female family members, whereas males prefer novelty. They are much more likely to move toward and play with elephants from other families, ideally males their own age or older. This does present dangers for male calves, though, and they are much more likely to become separated from their family or encounter predators, which certainly contributes to the greater likelihood of death that is recorded for male calves of all ages. On average, their more adventurous nature and greater nutritional requirements make it harder for elephant mothers to rear males successfully.

Young elephants love to play. Calves of both sexes spend a lot of time frolicking around and exploring, and anyone that's been lucky enough to spend time with them knows that they can be very entertaining to watch. They love a good water fight and will often spray each other at the river for a bit of fun. Sometimes they play on their own, practising physical skills such as how to use their trunks effectively, which often seems to involve waving it round and round their heads like a ribbon; or playing with objects they find such as sticks, logs or stones – carrying them, balancing them on their heads or trunks, or scratching with them. Sometimes they play with other animals; chasing and generally tormenting anything in their path, be it egret, warthog or wildebeest. And best of all, they play with other elephants.

Whenever I'm passing through Nairobi in Kenya, I always make sure to visit the David Sheldrick Elephant Orphanage, where dozens of rescued babies are looked after. It makes for a heart-melting experience. Every afternoon they come charging from the forest enclosure to be fed milk from bottles by the wildlife staff, but on the way the little elephants can't help but mess around; tripping each other up on purpose, or rolling down the hill causing a pile-up. I even watched one try to climb onto the other's head, just for fun. Once the milking was over, one particularly mischievous boy decided that my face would make for great target practice, picking up a piece of his poo and throwing it right at me, much to his human keeper's entertainment.

Play is an important source of learning and experience for calves and has long-term consequences for the development of their social and physical skills. For females, social play is one way to enhance their family relationships and to practise their mothering skills; for males, it is a way to begin building their social network with individuals that they will spend much of their adult life being with. But when scientists looked at data that followed a group of Amboseli elephants from birth to middle age, it became apparent exactly how important play is. Amazingly, they found that more playful calves lived longer as adults. Playing is not simply fun, it can increase an elephant's lifespan.

One of the main ways that play may help the calves is it allows them to learn how to communicate with other elephants. Elephants are constantly communicating with others, using visual gestures and vocal calls, as well as scent cues in urine and other secretions. With their large ears that can be moved, flapped, folded and wiggled, and their incredibly mobile and dextrous trunks, elephants can form a huge number of postures. Over a hundred gestures, postures and behaviours have been described that elephants perform and seem to have a meaning for those elephants who observe them – and for us watching, too, if we pay close enough attention.

These gestures vary from the subtle *ear fold*, which signifies anger or aggression, to the overt and obvious *let's go* sequence displayed by individuals who are trying to lead or direct the group in a particular direction. This latter action involves an individual standing on the periphery of the group, facing the direction in which they want to travel, perhaps with a front foot lifted or in front of the other, head up, and making a deep, low rumbling call every minute or so; looking back over the shoulder to check on the activity of those behind. As soon as the lead individual, who is usually the matriarch, has the attention of those behind, they will move off in the indicated direction, checking that the others are moving, too.

There are many gestures in an elephant's repertoire, and these can indicate a range of emotions that show everything from aggression and apprehension to friendliness, joy, and of course, their sexual intentions. However, we still know precious little about how exactly these visual signals work and the importance of the part they play in communication.

Given the way in which elephants' social lives operate and the fact that they often go wandering off alone, they must clearly find some way of keeping in touch with others whom they cannot see. Elephants are known to leave 'messages' in the form of scent cues via secretions from the temporal glands, as well as in their faeces and urine. But equally intriguing are the noises they emit, and scientists are beginning to explore the question of whether elephants can in fact 'talk' to each other with the same levels of sophistication as other big-brained mammals such as chimpanzees and dolphins.

Elephants can make different kinds of vocalisations. They're probably most famous for their trumpeting, a sound created by blasting air through the trunk. But even more impressive is their ability to generate deep rumbles, which resonate through the larynx all the way up to the nasal passages of the skull. Where trumpet calls are emotional sounds – often used by calves when playing, or when an adult feels threatened or in distress – rumbles seem to be given to maintain and reinforce social bonds, as well as to communicate intent.

One particular type of rumble vocalisation – contact rumbles – appears to be used for long-distance communication. Research in Amboseli has shown that female elephants are familiar with, and can discriminate between, the contact rumbles made by a mind-boggling more than 100 other elephants! And they can even recognise the caller from sounds heard at distances of up to two and a half kilometres away. Males can also discriminate between the contact rumbles of familiar and unfamiliar females, moving toward the unfamiliar females as part of their constant search for new mates.

These contact rumbles are low-frequency calls that have a deep, resonating sound. This kind of rumble can also be transmitted seismically – through or along the surface of the ground. Studies conducted in Kenya and Namibia suggest that elephants may be able to pick up these sounds through vibrations in the ground using sensors in their feet and trunks. Some scientists have even speculated that these rumbling communications might be transmitted over distances in excess of twenty kilometres! If this is true, 'messages' might be passed from one herd to another over vast distances, which might explain how these enigmatic animals can 'sense' danger quickly from far away.

Other sounds made inadvertently by elephants or from different sources could potentially also be detected in the same way. For example, a group of elephants rapidly running away from danger would create considerable seismic vibrations, as perhaps would distant thunder, or even tsunamis, which may explain the observation of Asian elephants running to high ground long before the 2004 Indian Ocean tsunami hit land.

One filmmaker describes seeing a family walking steadily across a dry lakebed in Amboseli. Suddenly, all the elephants stopped and stood completely still, with several resting their trunks on the ground – all looked as if they were in some kind of trance state of heightened concentration. After a few minutes, they relaxed, but all walked off in a direction at ninety degrees to their original path. Detecting distant sounds – and recognising something salient in them – is surely the most likely explanation for such sequences that are observed fairly commonly in elephants, but we still have a lot to learn before we can understand these almost telepathic abilities.

Until recently, it was thought that male elephants were fairly solitary creatures, but studies in Namibia, Botswana and South Africa have shown that is really not the case. Male social life falls into distinct periods. Male calves are born into female groups and remain with their family until adolescence, then when they are around fourteen years of age, these males start to move further away from their female kin to begin the process of becoming fully independent and establishing themselves in male society.

Gaining independence is a gradual process, taking place over several months. It was believed that independence was forced upon teenage males by their increasingly intolerant family, but – as ever with elephants – it seems that the individual's experience, maturity, personality, and relationships with his family and wider society all impact the decision to leave: a boy can't be rushed.

As they grow in confidence, young males explore wider areas in search of food, water and companions, and as time passes, they prefer to be close to other males and start to establish new friendship 'gangs' – bachelor groups. These young males can often be seen testing their strength against their peers, in friendly and not-always-quite-so-friendly sparring interactions: play fights can sometimes turn ugly.

The play fights and sparring of childhood and adolescence can determine dominance hierarchies that shape much of adult male life. Dominance is generally based on size – and therefore age – and strength. Older, larger males will take priority over subordinate males for access to food or water, and low-level aggression is not uncommon between males. But dominance hierarchies are most apparent when the stakes are high, that is, when resources are scarce.

In unusually wet years when water is plentiful, males in Etosha National Park, Namibia, display a lot of greedy pushing and shoving to get to watering holes. In drought years, however, the same males queue politely, waiting their turn to drink after more dominant individuals. They seem to realise that pushing ahead when the resource is so precious could result in *serious* fighting, and that is simply not worth the risk.

As males establish their place in society, they tend to live in areas that can be slightly removed from those in which family groups roam. In these bull areas, males form groups with others that can range in number from only two or three males together, to large groups of twenty or more. Some males are very gregarious, spending much of their time in the company of other males when not sexually active, while others prefer to spend more time by themselves. So just like humans, some are more extrovert and others more solitary. However, even large, old males typically have at least one long-term friend.

From late adolescence, keen young males will start attempting to pursue and mate with females. But these sexual interactions are rare and opportunistic. Then at some point usually in their mid to late twenties, adult males begin to enter their annual 'musth' phases.*

During these periods, males become considerably more aggressive, eating little yet wandering far and wide in search of females who might be receptive to their advances. A male's behaviour changes drastically between his normal, non-musth life and the musth phases, moving from the low-level, social jostling of the bull areas, to competing fiercely with any other male in his path for exclusive access to females.

Males can detect if a female is receptive to mating by the chemical signature of her urine, and also by the visual signals she gives him – essentially a flirty walk, wiggling her hips, and with a head movement that is akin to a wink and a hair flick. Musth males will chase non-musth males away from families that contain females giving off such signals. Musth status trumps all, so non-musth males always concede, even if they are normally higher up the pecking order.

---

* Musth was first noted and described in African elephants in 1981, by Joyce Poole and Cynthia Moss – the wisest matriarchs of all in the elephant-researcher family. Until they figured out what it was, they genuinely thought the males that they were seeing wandering alone, with no interest in food (and therefore with a declining body condition), and almost constantly dribbling urine, were sick with some mystery illness. In fact, they even jokingly called this 'illness' Green Penis Disease, based on the distinctly unusual colour they also noticed at these times.

That means that smaller, younger males who are in musth will dominate larger males who are not in musth. And if two musth males meet and neither gives in, terrible fights can break out that may last for many hours. There will be a lot of inevitable sizing up and squaring off, but when the real clashes come, they can be so prolonged and fierce that it's not unheard of for one of the males to be mortally wounded. To walk away with broken tusks and a few bleeding injuries would be getting off lightly.

Interestingly, the musth periods of male associates in Amboseli do not tend to overlap. Perhaps they have evolved in such a way that they avoid having to clash with their friends. Or it could be that they are only friends with males whose musth period does not overlap with theirs, which is an equally fascinating idea. Also, males who have reached the sexual and social maturity of having an annual musth phase are more likely to stick to a specific bull area when they are not in musth. We don't know exactly why males stick to these areas, or how they choose them, but it seems likely that the security such places provide helps them to recover body condition lost during the long period when their body is ruled by testosterone.

But it would be wrong to think that only musth males leave bull areas to spend time with females in family groups. Adult males who are not in musth are still sometimes seen in the company of families. Maybe this occurs by chance or maybe some males simply want the company of a talkative, busy family for a while. Perhaps they simply get lonely wandering alone. At the moment, we don't know for sure.

However you look at it, musth is very strange. Not merely the face-value strangeness of a three-month testosterone surge, but male elephants do not begin musth cycles until their twenties, if not their thirties. And genetic paternity tests conducted at both

Amboseli and Samburu have shown that males aged between forty and fifty-five are by far the most likely to father calves. So larger, older musth males are the most reproductively active. Which may not sound very odd, until you consider the fact that the average life expectancy for male elephants is around twenty-seven years in Amboseli – and that is probably the least disturbed, best-protected elephant population on the continent. Male elephants have evolved to sire offspring at an age by which most of them will already be dead.

It is not only larger, older males that father calves, but specifically larger, older males who are in musth. To survive and fight for months on end whilst eating very little food and quickly walking huge distances means males who reach and survive through annual musth periods must be very fit indeed. Physically fit, mentally fit, but also genetically fit, with good genes that could be passed on to offspring.

For biologists, musth is therefore an 'honest signal' of fitness – that is a characteristic or trait related to the genetic quality of a male that cannot be faked – and female elephants probably prefer to mate with males who demonstrate this honest signal of quality.

Three-quarters of males do not live to the age of forty, yet that is the age when the surviving males can count more reliably on fathering calves. And from a female's perspective at least, that makes perfect sense. A forty-year old male will be significantly larger and heavier than a twenty-year-old one, so more able to withstand the physical ravages of musth, not to mention being more experienced in all aspects of life. She wants only the best for her calves, and mating with older males demonstrating this honest signal of their superiority is a good rule of thumb for obtaining the best genes.

Males who have not yet reached musth age, and males who are in the non-musth phase of the annual cycle, can and do mate and father calves, but not very often. The extreme competitiveness of musth males, and the fact that females prefer these dominant individuals, means that musth males have the greatest success as fathers. So just as it is for females, males must also live a long time to produce a lot of calves. Longevity is key in both male and female elephants.

One particularly wonderful thing about wild elephants is that they also seem to accept people as part of their society. Many animals might form relationships with people – dogs and horses are prime examples – but these are typically evident either in domestic animals, or wild animals that are held in captivity and that form a relationship based on dependence on human caregivers.

Relationships with truly wild animals that are not based on any kind of feeding or provisioning are much rarer, yet almost all the researchers who have spent decades following particular populations of elephants will report a 'special' association with one or two of the individuals they see most often. Not simply that the researcher has a favourite whom they particularly care about or enjoy watching, as I am sure happens with almost all field biologists, but genuine two-way relationships that both human and animal seem to take pleasure in.

In his book *The Elephant Whisperer*, Lawrence Anthony tells the story of how he adopted a herd of 'rogue' elephants, who were destined to be shot because of their destructive behaviour across South Africa's KwaZulu-Natal province. 'They were a difficult bunch, no question about it,' he reflects on meeting them, 'delinquents every one. But I could see a lot of good in them too. They'd had a tough time and were all scared and yet they were all looking after and protecting one another.'

Lawrence and his wife, Francoise, had been approached by an elephant-welfare organisation when it all became too much for the herd's previous owners. They had space on their game reserve in the wilderness of Zululand, which, against all odds, became a conservancy for this troublesome bunch. Lawrence really was their only hope. The couple adopted the seven elephants into their family, naming the matriarch Nana, just as his children call their grandmother.

The early days were challenging. Every morning the herd tried to force their way out of the enclosure, and every day Lawrence persisted and did all he could to persuade them not to behave in this way. However, he also made it clear that despite their actions, he still cared for them and they could trust him, which given their previous experience of ownership, he expected would take a long time.

He pleaded daily with Nana not to break the fence, knowing full well that she couldn't understand him, but hoping that his body language and the warming, soft tone of his voice was enough for her to comprehend. To keep them in their compound was crucial for their safety, because their adoption and relocation had become national news, meaning they could easily be at danger from poachers and local tribes, who resented them being on 'their' land.

Over the months and years, Lawrence developed a very special bond with the animals and became convinced of their ability to express empathy, not only with other elephants but also with humans, and during this time he grew especially close to the matriarch of the herd.

'One morning, instead of trying to break down the fence, she just stood there. Then she put her trunk through the fence towards me. I knew she wanted to touch me – elephants are tremendously tactile, they use touch all the time to show concern and love. That was a turning point.'

Once Lawrence was accepted by Nana, the other elephants quickly followed suit, which proved lifesaving for him and Francoise when they inadvertently found themselves between the feistiest mother, Frankie, and her babies. She charged at the couple, only breaking off when she was seconds from crushing them. Had Nana not shown Frankie that she could trust them, they might never have survived.

The most astonishing and emotional tale of all – which really encapsulates the shared respect between the matriarch and Lawrence – is when Nana gave birth to her son Mvula. She ambled forward out of the bush, only a couple of days after the birth, with her new baby in tow, and presented him to the man she regarded as a close kinsman. A few years later, Lawrence repaid the gesture with his newborn granddaughter. In the same way that elephants celebrate a new member of their own herd, they shot their trunks up in the air to smell the scent of the baby and trumpeted in joyous unison.

For a wild animal to act with anything other than distant tolerance of a human – particularly one who is present a lot, but who never feeds them or gets directly involved in the animal's activities – is fascinating. For the animal to call to the human, to greet them as they do others in their social network, and present their calves to them, is truly remarkable. And that's not an isolated incident.

Probably the most enduring and endearing friendship between researcher and elephant was that between Cynthia Moss and a huge matriarch called Echo. Cynthia, who founded the Amboseli Elephant Research Project in the 1970s, named Echo early on in her research. For many years they encountered each other almost every day and spent months together as Cynthia filmed the herd in their natural environment.

Over time, Echo began to greet Cynthia in the vehicle as if she was a family member; and would even use the safari truck as an ally in potentially dangerous encounters with other elephants or lions; like she would rely on a sister. As with Anthony Lawrence and his matriarch Nana, Echo always came and presented her new calves, grand-calves, and even great-grand-calves to Cynthia – but never anyone else. If Cynthia was driving an unknown person in her vehicle, Echo could sense the stranger and would approach the car with much more caution.

It's clear that elephants are highly social and complicated creatures, which begs the question of exactly how intelligent are they? We know they have big brains and amazing ways of communication, but how does their mind work?

6

A Curious Mind

Hidden away on the banks of the Chobe river in Botswana is the headquarters of a charity called Elephants Without Borders. In the wooded grounds is a little orphanage where young elephants are being looked after. When I visited in June 2019, there were three calves called Panda, Molelo and Tuli.

Panda had been discovered alone in a farmer's field, and brought to the safety of the orphanage. Amazingly, despite her young age, she seemed to flourish as instinct kicked in and she assumed responsibility as matriarch of the trio. Molelo, aged only four or five, had been rescued from a bush fire and had been treated for burns on his feet and stomach. He was content to follow his new sister around, despite being the bigger of the two. And then there was Tuli, the smallest, whose parents had been killed by local villagers who were scared of losing their crops to the animals.

Despite the fact that they'd all suffered at the hands of humans, these three beautiful creatures were now bonding well and learning how to get along. It was the policy of the orphanage not to interfere with their upbringing other than to provide food. So the three elephants had to muddle along and learn from each other. As I wandered through the trees, bordered by the river, I watched as they frolicked and played with each other – and me.

Tuli would come over even though I didn't have any food for him, and would stretch out his hairy trunk and try to hold my hand, or slobber in my ear. He was clearly intrigued at his visitor and wanted to know everything about me. And as I discovered, there was no such thing as saying no. If I backed off, Tuli would follow. Molelo, the biggest, was more interested in his environment – he seemed to spend most of his time examining the natural world, picking up sticks and testing his weight against trees to see if he could break them.

But the most remarkable thing I saw there was when the elephants had finished their daily walk and knew that it was feeding time, and the three of them charged towards the gate of the compound.

Molelo barged up to it first and then, frustrated by the barrier, proceeded to headbutt the metal door several times as little Tulli watched on. But Panda, the girl–matriarch, clearly the more resourceful and perhaps intelligent of the bunch, approached with a more thoughtful tactic. She used her trunk to flip up the latch and pull on the bolt, sliding the door on its rails until it was wide open. Molelo seemed to give a scowl at his smarter companion, and not being much of a gentleman, charged forward first into the feeding grounds.

Not all elephants are as rude as Molelo, though, and I've seen plenty of examples of how they appear to show acts of consideration and thoughtfulness for others.

I remember watching two young males play-fighting in Kenya. One of the bulls was crippled, so the other got down on his knees to play, staying like that for the duration of the game, literally to level the field. Such observations are not rare, and experts have noted similar things watching different elephant populations right across Africa. Elephants seem to understand what others need, and what they can do to help.

But can that really be the case? Can an animal really act in ways that look to be so, well, human? We know that the social lives of male and female elephants are very complicated, with long-lasting and intimate relationships between individuals. And we keep coming back to the long lifespan of elephants, suggesting that experiences gathered over a lifetime make both male and female elephants better at producing and raising calves – more successful in a biological sense.

That long lifespan – including the extended childhood of elephants – not only allows for their immense physical growth, it also gives individuals a long time to learn about the complex social and ecological world that they inhabit. To learn *how* to be an elephant. And there a lot to learn.

For calves, much of the learning about how to behave and communicate is likely done through play, as it is with us. But a lot of knowledge is also acquired from being around and watching what other, older elephants do. This kind of learning from being with, observing, and copying others is termed 'social learning', and is hugely important for many animals, from ants and fish to chimpanzees and humans.

Social learning can take many forms and works through various mechanisms, but it is generally an efficient way to learn trustworthy information about what to eat, how to avoid being eaten, how to communicate, who to be friends with, who to mate with, and how best to raise a calf. We don't yet have much direct evidence that elephants learn in this way, but intuitively, it seems this must be the case.

Teaching is one very particular and special form of social learning. In humans, it is reliant on the teacher understanding

Elephants and men – cave paintings and petroglyphs at Boumediene, Tassili n'Ajjer national park, Algeria.

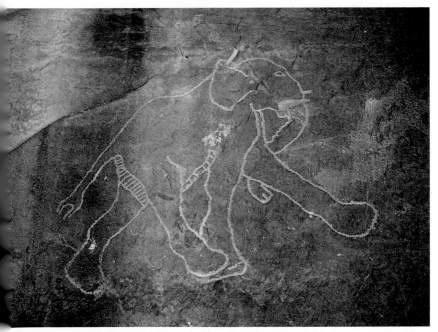

Portrait of a prehistoric elephant. Tadrart Mountains, Libya.

Hannibal in Italy – fresco, Capitoline Museum.

Ileophans sunt elephante. qo arabic alehy.

A medieval Italian drawing of a elephant c.1440. It appears like th Italians had forgotten what elephant looked like by the Middle Age

Despite the odd proportions, this medieval manuscript clearly shows how elephants remained in the European consciousness long after they disappeared.

A manatee. Part of the Afrotheria group of mammals and closely related to the elephant.

Rock hyrax (*Procavia capensis*), also known as the Cape hyrax, another Afrotheria mammal and the closest living relative to the elephant.

African elephants being trained at Garamba National Park in Zaire (Now DRC), 1972.

Jumbo – The famous elephant held in Regent's Park Zoo c.1880.
He also travelled across the United States in a circus act.

Jumbo, giving passenger rides in London.

A mature elephant feeding in the swamps of the Okavango delta. When elephants get old their teeth wear down, meaning they need to stay close to soft vegetation. Often, they die in these areas giving rise to the myth of the elephant graveyard.

n African elephant in s prime standing at 3.4 etres tall. Botswana.

A breeding herd of females wallowing in a mud pool to help thermoregulation.

An old Matriarch leads her family, Amboseli, Kenya.

Elephants are born swimmers and use their trunks to breath.

Youngsters sniffing the air. Trunks play an important part of an elephant's social life and can detect smells five times better than humans.

A breeding pair clearly shows sexual dimorphism. The male in the foreground is larger, with bigger tusks than his mate. Her head is more pronounced.

Elephant footprints in the Makgadikgadi Pan. Young bulls lead the way, exploring new routes in search of food and water.

Two young males play-fighting under the shadow of Mt. Kilimanjaro.

A young mother
with a new
calf flanked by
her sister.

A mother with her older
calf. Family is important
to elephants.

precisely what the learner knows and does not know, and how to communicate the information that the learner needs. Human teaching is not restricted to school education – we all explicitly teach each other a lot of the time. It is probably not that long ago that you showed someone, or were shown, how to do something on your phone or computer, for example.

In animals, this kind of social learning is rare, because it relies on understanding the knowledge of the pupil, which is a complex and rare capacity, as we shall see later. But it is possible that young, naive female elephants are taught about sex – specifically who to mate with – by older, more experienced females.

After analysing records of females in Amboseli faking signs of 'oestrus' (characterised by the flirty walk) when they could not have been sexually receptive, researchers concluded that the most likely explanation for the simulation was to show young females how to behave and to whom they should direct these signals. Only older females simulated the signs of sexual receptivity in this way, and only when they were in the presence of young, naive females, who were becoming sexually receptive for the first time.

It is important for all females to mate with the 'best' males, as we have already talked about, but especially so for the youngest and smallest females. Having a large male climb on to their back to mate can be extremely dangerous for females. So much so that one adult female in Amboseli walks with a permanent limp, and I was told by researchers that they think this is a result of a mating injury.

But somewhat surprisingly, the younger, smaller males may be the most dangerous to mate with, as they do not yet have enough strength to support their own body weight, and so lean more heavily on the female as they mount her. Larger males are more adept at standing on their back legs. Fascinatingly, it seems that smaller females have to be taught this vital information and encouraged to mate with the much more intimidating, but ultimately a safer bet, larger males. Surely that's as clear a demonstration of social teaching as you can find?

Learning which actions or responses are appropriate in what circumstances is an important part of learning how to be a successful adult elephant. The same could be said for us humans with social cues. It's been shown that the older an elephant, the better it can recognise familiar sounds. Older matriarchs have greater knowledge of the social network around them than younger family leaders, recognising a greater number of other families and accurately identifying unfamiliar groups that could present a social threat. Also, by not wasting time responding inappropriately, groups with older matriarchs have more time and energy to dedicate to producing and successfully rearing calves. The greater social-network knowledge of older, wiser females is one more reason why families with older matriarchs are more successful.

Matriarchs don't solely offer knowledge about social networks to their families. They also learn and share information about the ever-present risks in the savannah environment. I've seen it a couple of times, when younger female elephants have heard a number of lions roaring; they would react, become agitated, and look to the matriarch for guidance. However, they would often under-react to hearing the roars of male lions, perhaps because they thought there were fewer of them; apparently not fully aware of the greater danger that male lions can pose. An adult male lion can weigh up to 50 per cent more than an adult female and they are more likely to engage in riskier hunting strategies, including tackling larger prey such as elephant babies.

Interestingly, older matriarchs consistently show better and more adept decision-making in response to the threat of a nearby male lion, bunching their family together in a tight defensive formation, with calves safeguarded in the very centre of the group. In real-life encounters with lions, as I saw in the watering holes and pans of Botswana, such reactions are likely to protect and prolong the lives of vulnerable members of the family.

Families with older matriarchs are also better able to survive periods of drought, because these females have knowledge of where water can be found even in extreme circumstances. Older matriarchs lead their families over a much greater area during periods of drought than younger matriarchs who lack the knowledge, and more calves survive droughts in families with older leaders. It's staggering to think how much wisdom and knowledge they must accrue and how important it is for the herd.

In fact, directing their family to safe areas is the key job for matriarchs. Safe areas with enough food and water to sustain the family, that is. Better yet, safe areas with food and water that also allow meetings with other elephants – like mothers taking their young children to the park for play dates to learn how to socialise. Experienced matriarchs know that leading their family towards social hotspots, to encourage and facilitate friendly interactions with others, can be equally as important as leading them to water or away from lions. The best decision a younger matriarch can often make is to follow another family led by an older female, so that they too can benefit from her accumulated wisdom.

It is not only families that benefit from the leadership of a wiser elder. Independent males who no longer live with their family also need guidance, learning appropriate behaviour from their seniors. We know that older males in musth wander further and faster looking for receptive females, and the benefits of this strategy are presumably one of the key behaviours that males learn as they age. Moreover, males of all ages tend to prefer being near an older male, over thirty-five years old.

This proximity to older elephants gives younger males the opportunity to observe and learn how to behave from more experienced individuals. As females need to learn how to be good mothers and family members from older females, males also have a great deal to learn from their elders – about the male hierarchy and acceptable social behaviour; about the best spots to feed, drink and hang out; and how to find, attract and keep females to mate with.

It is clear that for elephants, much like people, knowledge accrues with age, and elephant scientists often repeat the phrase that older elephants are 'repositories of knowledge'. However, having knowledge and experience is all very well for an individual, but it doesn't mean much in a society if that knowledge is not shared and passed on. We assume that it is acquired by elephants through some of the various forms of social learning. This sets up the intriguing possibility that elephants may show cultural differences, as humans do, in what they learn and how they behave.

Human culture, of course, is about so much more than what we learn from others. It is about how we teach, and keep ourselves and others accountable for what we have learnt; how we define moral and ethical codes of conduct that we expect our society to adhere to, punishing or excluding those who behave differently from us. It is also how we build on the acquired knowledge of previous generations to formulate new ideas, customs, trends or technologies. Few expect that any animals will show these kind of cultural traits – indeed, our acquired culture is probably the one thing that really sets us apart from other animals – but investigating proto-cultures, or at the very least behavioural traditions passed down through generations, is a hot topic in animal behaviour science right now.

Investigations of any such traditions in elephants are ongoing, but we now know that individual and family position within the wider social network is maintained according to custom, and that the paths elephants use to walk to regular resources such as watering holes are traditional. In fact, some of these ancient trails have been used for hundreds and possibly thousands of years.

We all know that elephants have impressive memories, but we also have evidence to suggest that elephants can learn new sounds and vocalisations – imitating each other, and even copying people. One elephant, who has been housed alone for many years in a zoo in South Korea, apparently mimics the verbal commands of his human keepers with considerable success. In the absence of other elephants, perhaps he is trying to communicate with the only animals that he does have regular contact with.

Similarly, a young male savannah elephant, raised in the David Sheldrick elephant orphanage in Nairobi and subsequently released to Tsavo National Park, was recorded imitating the engine roars of trucks that passed by on the busy Nairobi–Mombasa highway that cuts through Tsavo. Since he was first recorded copying this noise in the early 2000s, the habit has now spread to several other elephants.

We do not know if this kind of exact copying is how young elephants learn the repertoire of rumbles and trumpets they will need to function in society, or if it is how any other skills or trends are passed down through generations, but it does suggest that in their own way elephants can pass on information to others, as other intelligent mammals like chimpanzees and dolphins are known to do. Learning huge amounts of information about the world is all very well, but how do elephants, old or young, manage all this information to make appropriate social and ecological decisions? To answer that question, we must now turn to thinking specifically about minds and intelligence.

Elephants fall very naturally into the group of animals that many of us find 'special', like dolphins and the great apes – chimpanzees, gorillas and orangutans. They are all fairly long-lived, with long periods of childhood spent living in and learning from large and complex social groups, and all have large brains – both when compared to their own body size, and in absolute terms (as compared to other animals). We are not often surprised when we read media articles reporting 'intelligent' behaviour in these species, because it seems natural for us to assume that animals that live like us, might also have some kind of intelligence like us.

But what is intelligence? Can human and animal 'intelligence', whatever it may be, truly be compared? Are some species really cleverer than others? And if elephants are intelligent (and therefore special?), what, if anything, does that mean for their place in a world that we dominate?

The term intelligence is heavily loaded and ill defined, even when thinking only about humans. Some people might equate intellect to success in school exams or university degrees, others may think of it as speed of thought – catching on to ideas quickly or having a lightning response to a cutting remark; for others still, violin prodigies or artists are clever.

The point is, though we all have opinions about who is intelligent, and rarely show qualms about describing some people as 'clever' and others as less so, intelligence is difficult to define. To sidestep these difficulties, many scientists adopt a more technical-sounding word in the hope of achieving greater transparency about what they are researching. And that word is 'cognition'. Most psychologists and biologists who study the mind or 'intelligence' of animals would typically say they study animal cognition.

At its broadest, cognition is about processing information. The term came to prominence with the advent of computing in the mid-twentieth century, and cognitive psychologists tend to agree that it refers to acquiring, storing, retrieving and processing information about the world.

Studies of human and animal (and indeed computer) cognition ask what information the mind (or machine) in question is *representing*. What information does a particular individual notice about the world and themselves, how do they store and remember this information, and how and why do they use it. Essentially, we are asking what information this individual requires to function.

So, what information do elephants need about their world, and how does that information help them?

Whole books are dedicated to the topic of cognition in species like domestic dogs and chimpanzees, based upon literally thousands of research papers produced by hundreds of scientists around the globe who are studying these animals. For elephants, such books do not exist, simply because the science is not yet there. Whilst many scientists research elephant behaviour and ecology, relatively little is known about elephant cognition because – quite frankly – it is very difficult to conduct studies exploring elephant minds.

Most studies of animal cognition have traditionally been conducted using captive individuals that can be approached easily and trained, or rewarded for interacting with the experimental apparatus the scientist has arranged. But whilst there are arguably too many elephants held in captivity, there are few African elephants that live in the kind of captive conditions that make such testing viable. And designing experiments that can be conducted with wild elephants is rather challenging, limited by the fact that we cannot actively encourage or reward participation, we cannot control who takes part in the experiment, nor can we use any apparatus or design that would require training.

As such, studies of elephant cognition are still in their infancy, but this slowly blossoming field has nevertheless revealed some interesting results so far.

Imagine you are visiting family for the day – your sister or aunt, for example, and her numerous children – with your own immediate family in tow. Let's say around twenty people are there. It is loud and chaotic and boisterous, with gentle teasing, gossip and excited conversations. A walk outside seems like a good idea, so eventually you manage to corral the troops and get everyone into the street. Simple.

Now all you have to do is keep an eye on them all as they move along, that's your job. The only problem is that some are running, some are walking and some are downright dawdling. You have to keep track of who wants to go where: who wants to cross the road to follow a butterfly, or to stop in the newsagent's, or put a note through a neighbour's postbox. You hope everyone knows where you are aiming for. But it may well be that until you do all meet up again, and everyone is tied down to a park bench with an ice cream shoved in their mouth, this will feel less like a relaxing walk and more like a nightmarish intelligence test.

Well, it turns out (not to make you feel worse about managing your own family walks), this is the kind of test that elephants can pass with flying colours. Elephants, as we discovered in the last chapter, can differentiate between dozens of other elephants simply by the noises and rumbles they make. It turns out, as I learned from Lucy Bates, who had been working alongside the elephant expert Joyce Poole in Amboseli, that they can also recognise each other from the scent in their urine. And that is how the matriarchs, and other members of the herd, keep tabs on who is where.

Lucy Bates explained some of her own observations in the field in 2006: 'Joyce looked up, watching an adolescent female elephant sniffing the puddle of pee there from another elephant. The thought suddenly came to mind, why does she care? I like to think Joyce and I looked at each other, having the idea at the same time: let's test what they know about each other (their knowledge acquisition, storage, retrieval, and processing) from sniffing each other's pee.'

When Lucy returned to Amboseli the following year, she set about collecting elephant urine. Lots of it. By collecting urine from known individuals, and presenting it on the path of elephant families and recording their reactions, she was able to discover that elephants can use scent cues in urine to recognise specific individuals from their family.

'In one part of the study, we collected urine from a family member who was walking at the back of the group, quickly drove forward with the sample, and laid it out on the path ahead of the front of the group. The first female who reached this sample showed intense interest – and surprise – in the urine. We concluded that the female's surprise was because she not only recognised *who* the urine was from, but that the sample should not have been there in front of her, because the female who made it was actually walking behind her. How could that female have just urinated on the ground in front when she was behind the whole time?

'We resolved that elephants are able to keep track of multiple family members at the same time, knowing where these others are in relation to themselves. They can *recognise* and differentiate the individuals in their family, *remember* which individuals are with them, and *recall* specifically where each individual is currently.'

Recognition of individuals is not uncommon in mammals that live in cohesive social groups, but the ability to keep track of where so many of these individuals are at any one time has not been documented in other animals. The social organisation of elephant society, and the fact that they will go wandering off on their own for extended periods, makes managing their associations especially complex, and to keep track of individuals like this requires impressive computational abilities.

Specifically, it requires *working memory* – the term cognitive psychologists use for the brain function that allows you to keep a limited amount of information temporarily to hand for immediate use, like the RAM memory of a computer.

The elephants Lucy tested were keeping track of at least seventeen individuals at once, based on the size of their family groups. An average adult human can readily keep about seven or eight bits of information (digits in a phone number, location of children, that sort of thing) in mind. The working memory capacity of elephants is, it seems, considerably greater than that of the average person.

Recognition and remembering is based on perceiving and categorising the details of things, and unsurprisingly elephants are known to have excellent categorisation skills. For example, something that is truly remarkable is that they can recognise and categorise ethnic subgroups of humans based simply on how those ethnic groups smell or sound!

Lucy told me a bit more about her findings: 'We presented Amboseli elephants with red-coloured cloths that had been worn by either a Maasai warrior or a Kamba man, or a red-coloured cloth that had not been worn by anyone. Maasai warriors occasionally spear elephants around Amboseli, whereas the agricultural Kamba rarely pose any mortal threat to elephants in the area. Elephants were able to determine the difference between cloths worn by Maasai and Kamba men based on their scent only.'

When detecting cloths worn by Maasai, the elephants rapidly fled the area, usually heading to an area of greater cover such as long grass. Cloths worn by Kamba men caused only mild alarm in comparison; the elephants usually did move away, but at a much slower speed, for a much shorter distance, and relaxing again much quicker.

Two other elephant scientists, Karen McComb and Graeme Shannon, went on to show that elephants in Amboseli can also discriminate between Maasai and Kamba men based on their voices. They played recordings of men saying, 'Look, look, over there, a group of elephants is coming,' in their native language (Ma or Kikamba), and again found the most heightened response to the voices and language of Maasai men. The reaction was specific to the age and sex of the Maasai speaker, with less fear shown toward Maasai women and boys than adult men.

Interestingly, the reactions observed to even adult male Maasai speakers were not as extreme as in the scent tests, and all the researchers agreed that this is probably because Maasai warriors were unlikely to speak out loud when hunting. Thus, hearing Maasai voices means that they are probably not intending to threaten the elephants, whereas smelling silent warriors could occur when the warriors are trying to conceal or hide themselves, which would of course end badly for the elephants.

If working memory is like the RAM of a computer, long-term memory is the data storage capacity. And impressive long-term storage and retrieval of information has been recorded both experimentally and anecdotally in elephants. Rather touchingly, Karen McComb showed that elephants in Amboseli still recognised the contact calls of a relative nearly two years after the female died.

Several reports exist of captive elephants, who had lived together as youngsters and then been separated for decades, still remembering and recognising each other years later, showing intense excitement when they were reunited.

For example, Shirley and Jenny are two elephants who spent the winter of 1973 together in an American travelling circus. Jenny was less than two years old at the time, and Shirley was around twenty-five. After the brief season together, they were each sold to different circuses, and subsequently to zoos across the US. In 1996, Jenny was handed to the Elephant Sanctuary in Tennessee, where her life-long injuries were finally treated and she could begin to associate again with other elephants.

Then, in 1999, another much older female elephant arrived at the Sanctuary, and to everyone's amazement, the two elephants engaged in a highly emotional, exclusive, and prolonged greeting. The second female turned out to be Shirley, and only after much digging did the Sanctuary staff discover that these two elephants had indeed lived together for those few brief months twenty-five years earlier, when Jenny was still a very young calf. I challenge you not to well up when watching clips of the reunion on YouTube.

It's not only social information that elephants can remember over long periods. The ability of mature matriarchs to lead their family to distant water sources or feeding areas, decades after having last been there, also relies on long-term memory. Ranging behaviour can also be limited by their memories. At several sites in southern Africa, researchers have noticed elephants may take years to cross into land that has been reclaimed into wild habitat. The elephants will walk up a now imaginary line, where there used to be a boundary fence, but not walk over this 'frontier'. That 'elephants never forget' does indeed appear to be true, even though their memories may not always be an advantage.

Elephants are clearly very good at noticing, learning about and classifying relevant information from their environment, and this information can be stored, retained, and recalled over the long term. But is that all elephants can do? Human intelligence is also founded on impressive learning and memory, but there is so much more to us than simply noticing and remembering things.

Our ability to compute and predict is surely part of what makes us human. We can *solve problems* with only minimal experience or information, we can understand what *others are feeling* based on their subtle expressions and our own experiences of similar situations, and we can work out *what others know*, believe, or might do next using our understanding of complex situations. Whilst these abilities – variously termed 'insight', 'empathy' and 'Theory of Mind', to psychologists – are profoundly developed in humans, it seems only a very select group of other animals possess any similar skills. Notably, it currently looks like the great apes (our closest evolutionary cousins) and a number of relatively large-brained birds, such as rooks and crows, share a few of these abilities to some extent. Domestic dogs have extraordinary social understanding, and there is slight but tantalising evidence that some dolphin species also possess amazing abilities.

So, do elephants also appear on this list? The answer, in short, is maybe. There isn't a massive amount of data to prove it, but what does exist certainly suggests that they do.

In an analysis of behavioural data taken from four decades of observations in Amboseli, elephants were clearly and frequently seen to display empathy in the form of protecting or comforting the vulnerable, and actively helping those who are in difficulty. Watch any elephant group for long enough, and you will see them forming defensive circles around younger individuals when they detect danger, or protectively laying a trunk over the more vulnerable.

Mothers and sisters may be seen adjusting vegetation or the position of a sleeping calf to apparently make them more comfortable. Able-bodied elephants will follow, remain with, and assist weak or injured individuals, perhaps helping them to stand and walk, or even passing them food. And larger individuals will always help calves out of rivers or ditches with steep, slippery banks, pulling or pushing with their trunks, leveraging with their tusks, or simply headbutting them out of danger.

There are many reports and videos on the internet of elephant mothers remaining with calves who are stuck in mud. And if humans intervene to assist the calf, the mother or remaining family often appear to step aside, apparently understanding that the people are there to help. Sometimes, the adult elephants are undoubtedly pushed aside by Land Rovers rather than stepping back willingly, but they rarely attack the vehicle moving them, instead walking away and quietly waiting. 'Caring' and 'understanding' are the two words I most commonly think of when watching elephant families go about their normal daily life. Everything seems thought out, and takes other elephants into consideration.

Whatever we make of the anecdotes, there is clear scientific evidence of elephants forming coalitions and alliances, solving problems cooperatively, and understanding the physical competence, emotional state, and intended goals of others. In short, elephants know that others are beings who can also act, feel and want things.

Understanding that others have their own emotional and mental states – the foundation of a complex suite of abilities termed Theory of Mind – may sound commonplace, but it really is not. For some, it remains controversial to claim that any animal except humans possesses any such understanding, although many psychologists would now agree that a number of other species do possess varying degrees of Theory-of-Mind-type skills.

Brain research conducted in the past decade supports the idea that elephants share some empathic and social skills with humans. A type of brain cell called a 'von Economo neuron' (after the man who first described them in 1926) has been discovered in specific areas of the brain of humans, great apes (but not monkeys or other primates), orca, humpback, fin and sperm whales, and African and Asian elephants – all species with large brains and large social networks.

These neurons appear to be involved in social and emotional awareness, empathy, and allowing fast decision-making in complex social situations. The fact that so few animals possess them is intriguing. But related to social awareness is an even more controversial ability: being consciously aware of one's *own* mind.

For some philosophers, self-awareness is an intractable problem. How can we ever be sure that any other person has a mind like our own, that they experience the world in the same way we do? Scientists tend to accept the argument by analogy when it comes to the minds of other people; you look like me, behave like me, and use language like I do, so you must have a mind and be aware of it like I am.

But scientists show much less agreement on the argument by analogy when it comes to animal minds, because how much 'like me' does an animal need to be, before we can accept it has a mind and is conscious? For some, until that animal uses language and can give us a running commentary on its thoughts, we should never accept it is conscious or self-aware. For others, if the animal demonstrates aspects of Theory of Mind such as advanced empathy, that should be enough. Others still take the stance that, until proven otherwise, we should assume that all creatures are conscious like us.

In an attempt to settle this debate, the 'mark test' was developed in the 1970s, to measure 'mirror self-recognition'. A coloured mark is surreptitiously placed on an animal's head, and the animal is then allowed to look in a mirror. If the animal uses the mirror to guide investigation of the mark on its own body, the animal must recognise the reflection it is seeing is itself, so the argument goes, and so it can be considered to be self-aware. Almost all animal species tested only respond to the reflection at best as though it were another individual, not themselves.

Apart from humans, the only species that reliably pass the mark test in the traditional sense are great apes, and – with some adjustments because they don't have arms and hands – bottle-nose dolphins and magpies. No, apparently not even your pet dog or cat really knows that it is looking at itself, whereas chimpanzees do naturally connect the reflection they see with their own body.

African elephants have, to my knowledge, never been exposed to the mark test. But captive Asian elephants have been tested, passing at about the same rate as chimpanzees. Although only a very small number of elephants have been tested, and we should not assume that because a few Asian elephants passed that African elephants would too, undoubtedly it sets up a very intriguing possibility. Elephants may well be self-aware.

This possibility is backed up by anecdotal observations noted by researchers studying elephants in the private game reserves bordering Kruger National Park in South Africa. They have noticed one particularly handsome, young male elephant, who likes to walk alongside the research vehicle for extended periods of time, constantly looking at his reflection in the windows. Perhaps even elephants can be vain!

Self-awareness, at least as measured by the mark test, is rare and remarkable within the animal kingdom. That is not to say that other animals, including your dog or cat, are not *sentient*. Other animals may well have feelings and emotions, and may recognise or even understand emotions in others, but in failing the mark test, it seems they do not experience a sense of self in the same way that we do, or chimpanzees do, or elephants might.

Thinking about empathy, minds and self-awareness leads us to thinking about an area of behaviour for which elephants are particularly famous – their reactions to death. Elephant grave-yards have been mythologised for years, but, much to my own disappointment, after having been brought up on the stories by my grandfather, the graveyards are merely that – myths. Elephants do not head to a graveyard to die, nor do other elephants trans-port the bones of dead friends to be buried in such an area.

There may be areas with higher concentrations of elephant bones, but these most probably occur because elephants who are frail and near death seek out and remain in areas close to fresh water and soft grass, so they can feed and drink more easily in their final days or weeks without having to wander far. Because such areas may be rare in arid savannahs, these swamps can, over generations, become magnets for dying elephants, resulting in a build-up of elephant bones.

So, any 'graveyards' are circumstantial, not intentional. But that does not diminish the reactions of other elephants when they encounter such bones, or carcasses of elephants who died more recently. Terms like 'mourning' and 'grief' are very loaded – obviously implying an awareness of death, and therefore an understanding of life. For this reason, animal scientists tend to avoid such terms for fear of suggesting knowledge in the animal species that we cannot be sure they have. But with elephants (and, interestingly, chimpanzees again), it becomes extremely difficult to describe their reactions to death without invoking such terms.

The death of one adult matriarch called Eleanor, from the Samburu population in Kenya, has been described in detail. As the female deteriorated, unrelated adult females showed a lot of interest in and compassion towards her; lifting her when she collapsed and trying to encourage her to stand and walk, remaining with her and vocalising and touching her for several hours once she was unable to stand anymore. Eleanor's own family was some distance away at this point, seemingly unaware that their matriarch had fallen. After her death the following morning, they remained near her body for seven hours, whilst other adult females gently investigated the carcass, sniffing and touching it with trunk and feet.

One day after Eleanor's death, another female, who had tried to help and lift her, returned to the area; this time standing still and silent by the body. Eleanor's own family were this time present, and her calf was seen to nuzzle her body, remaining in place even when other families came to investigate. Eleanor's family returned to her body several times over the following days, even when lions and other scavengers were in the area.

Even after the body decomposes, elephants recognise and interact with bones from other elephants rather than those of other animals. Buffalo or rhino skulls garner only a passing sniff, whereas elephant skulls or tusks are investigated intensively with lots of gentle touching and sniffing.

In her book *Elephant Memories*, the conservation pioneer Cynthia Moss recalls how she brought the jawbone of a recently deceased matriarch back to her camp. Several days later, the dead elephant's family happened to pass nearby and came into the camp to inspect the jaw. The animal that showed the most interest, lingering long after the others had moved on, was the matriarch's seven-year-old son.

Moss and her colleagues have followed up on such anecdotes with controlled experiments designed to explore this behaviour more systematically. When presented with three objects – a piece of wood, an elephant skull and a bit of ivory – elephants showed a marked interest in exploring the ivory and a clear preference in the skull over the wood.

We don't know exactly what this all means – what elephants understand or feel about death – but the fact that these kinds of reaction are seen in so few animal species beyond humans is certainly compelling.

We still have many more questions than answers about the minds of elephants. We can hardly begin to imagine the rich tapestry of how they communicate with each other, using chemical scents, vocalisations and vibrations, subtle visual gestures and postures. We know so little of what they understand about the world, or how they represent information and use it to solve problems. We don't know exactly how they learn from others, or what they really appreciate about the minds of others.

These questions remain unanswered at present. Scientists are trying to address them, one step at a time, but the steps may well be too slow. Our emerging attempts to probe what may be one of the most extraordinary brains on the planet are occurring at a time when those brains – and the bodies they are in – are being hounded and persecuted, possibly to extinction. Now it's time to explore where it all went wrong.

7

A Continent of Ghosts

At the start of the twentieth century, Africa's elephant population numbered anywhere between 3–12 million, but this was already a fraction of the tens of millions that likely existed only a few hundred years before. And now? Recent surveys have shown that there are only 415,000 African elephants left. The number of elephants in Africa has fallen by more than 90 per cent in little over one century. What on earth went wrong?

Many of the problems that elephants have faced in the past few hundred years are encapsulated in the example of what is now Addo Elephant National Park, in South Africa. Lying on the southern Cape coast close to Port Elizabeth, Addo was first inhabited by Khoesan people, then by the nomadic Xhosa tribe. That was until the Boers – white Dutch-descended farmers – arrived in the 1740s. The inevitable clashes between the three groups continued for almost one hundred years, until the Boer settlers finally drove the Xhosa and Khoesan out of the area. Throughout this time, elephant killing had been rife, with ivory trading being a major incentive behind the Boers' desire to control the land.

By the early 1900s, only a few isolated elephant groups remained in the area, perhaps as few as 130 animals – significantly down from the tens of thousands that had been present centuries before. Then in 1919, the professional hunter Major P.J. Pretorius was appointed by government officials to shoot all the remaining elephants in the area, in an attempt to end the angry conflicts with the Boer farmers surrounding Addo, and to allow more space for agriculture. Pretorius's actions resulted in the deaths of 114 elephants in thirteen months. Only sixteen elephants were said to be left alive.

Fortunately, public opinion changed somewhat after this mass slaughter, and by 1931 Addo was proclaimed a national park to provide sanctuary to the remaining elephants. Over subsequent years, more elephants were lost in continuing conflict with farmers, and at one point the herd shrunk to only eleven creatures. But, as awareness about the very real threat to the survival of the species hit headlines around the world, the population did begin to rise again. In 1954, an 'elephant-proof' fence was designed and built, enclosing and protecting the then twenty-two-strong elephant population by preventing them from moving into the surrounding farmland.

Today, the population of Addo is slightly over 600 elephants, and the area of the park available to the elephants has been expanded to accommodate their growing number and reduce the amount of damage they are causing to vegetation within their fenced area.

The elephant population has grown rapidly, despite the fact that its genetic diversity was low, as would be expected from having grown from a bottleneck of only eleven individuals. Low genetic diversity reduces the evolutionary adaptability of a population and can result in a greater incidence of inherited diseases – much like the inbreeding that blighted royal families in the past – so in a bid to add some variety, adult males were introduced from Kruger in the early 2000s to try to counteract this problem.

The Addo story is typical of the plight of elephants throughout the twentieth century and highlights many issues: population decline due to heavy hunting and poaching, loss of habitat and conflict with people over land use, and management or manipulation of populations by humans. But before we think about these issues in more detail, we have to first understand something about demographics and population biology.

The size of any population or species increases with births, and immigration. And populations decrease because of death, and emigration. Longevity, age distribution, sex ratios, age of sexual maturity, gestation and weaning all affect population sizes, and the *rate* at which the population may grow or contract. If everyone in a population lives longer, so there are fewer deaths each year, and births continue to occur, the population size will increase, as does the growth rate.

But if there are too few females or males, or more specifically, too few of breeding age, birth rate will decline and the growth rate will start to drop, which could ultimately lead to a decline in population size. And of course, if many individuals die prematurely or emigrate, at a rate that is higher than the rates of births and immigration, the population size decreases, and the growth rate contracts.

In short, the size, age and sex structure, and potential for reproduction and growth – the demographics – of a population are key factors influencing whether it is vulnerable to extinction.

By looking at demographic factors such as average lifespan, age and sex distributions, and the length of time between each birth among females, the theoretical maximum rate of increase for an elephant population was calculated as 7 per cent. This means that – under the very best conditions – a healthy population of elephants could increase in number by 7 per cent each year, on average.

So, in a population of 100 elephants with a normal social structure, optimum balance of reproductively active males and females, and natural rate of deaths, we could see an increase to 107 individuals, in one year. And the population will increase by another 7 per cent in the next and subsequent years, so that ten years later the population could be as large as 184 elephants.

Of course, few elephant populations meet these 'perfect' criteria (although Addo did), and so observed rates of increase – or population growth – tend to be lower, typically no more than a 6-per-cent increase per year in very good environmental conditions. And any population that is increasing at a rate that is higher than 7 per cent per year – as some have been recorded – is likely doing so because of immigration. That is, other elephants from outside are coming into the population, swelling the numbers.*

We can use these theoretical and observed figures of population growth rates to calculate how many elephants can die

---

* Interestingly, another situation that has seen growth rates exceed 7 per cent is the introduction of elephants to game reserves where new populations are being established (a common management approach in South Africa, for example). Rapid growth in these populations is facilitated by abundant resources and a young age structure, which leads to lower ages of sexual maturity and short inter-calving intervals.

before a population will start to shrink – the point at which we lose more elephants than are born each year. So, when we talk about elephant populations declining, it is a shorthand way of saying that the number of deaths (and emigrations) in the population is greater than the number of births (and immigrations). And of course, the starting size of the population is important in this calculation.

A population of 2,000 elephants can easily 'afford' to lose ten elephants a year – in fact, this is what we would expect to see based on average natural mortality patterns (at 0.5 per cent). But for a small, isolated population of only 100 elephants, the same loss of ten elephants in a year – or 10 per cent of the population – would be catastrophic, and way outside the sustainable number. Such losses in a small population would soon result in the extinction of that particular group.

It is therefore not surprising that the first areas to completely lose – or at least come close to losing – their elephants were at the northern and southern limits of their range, across North Africa (in the Middle Ages) and the South African Cape areas, including Addo (in the last century). Living in these areas was already difficult for elephants, given the habitats; arid in the north, rocky and exposed with a harsh climate in the south, so the population growth rates were probably low to start with. This means there was very little elasticity to withstand the onslaughts of intensive hunting that began with the arrival of Europeans in the 1500s, and the point at which hunting would have reached unsustainable levels would therefore have been much lower than in larger, more robust populations elsewhere in the continent.

The beginning of this book outlined how the Europeans arriving in Africa in the sixteenth century were the catalyst for subsequent population declines, setting off the scramble for ivory that continues to this day. Ivory had been coveted and used in many cultures around the world before this time, including some trade with Europe, 'Arabia' and Asia, but it was the early colonisers who have to take the credit for initially establishing the catastrophic levels of demand that persist. As early as the year 1670, John Ogilby wrote of the diminishing supply of ivory in Congo in his atlas *Africa*, presumably based on information he received from traders.

In the 200 years between 1500 and 1700, it is estimated that 100–200 tonnes of ivory were being exported from Africa per year. This figure exceeded 200 tonnes per year during the 1700s, and between 1800 and 1850, the figure was more like 400 tonnes per year. During the early stages of this mass export, the average tusk weight may have been as high as 12 kg. If we take this high figure – meaning one elephant could provide 24 kg of ivory – and an average export value of 150 tonnes per year, that gives an estimate of 6,250 elephants dying *per year* between 1500 and 1700, or 1.25 million elephants over that initial 200-year period.

That is a conservative estimate, using a high average tusk weight, and mid-point value for the number of tonnes of ivory being exported. Given that by the 1800s, average tusk weights had fallen considerably, a similar number of more than one million elephants were likely also slaughtered in the 1700s, and another one million again in the fifty years up to 1850. Considerably more than 3 million elephants were killed for their ivory in 350 years. But things got *really* bad in the second half of the nineteenth century. Between 1850 and 1900, approximately 700–800 tonnes of ivory were being exported from Africa every year.

Let's call that 750 tonnes, for fifty years, which equals 37,500 tonnes. Or 37.5 million kg. By this time, average tusk weight per shipment could be as low as 4 kg, but again let's be very conservative and use a figure of 8 kg, so each dead elephant could contribute 16 kg to that total. That means well over 2 million elephants were slaughtered for ivory in only fifty years, probably more, as tusk weights decreased.

Given that there is evidence (from tusk circumference) that females were increasingly being shot in some locations during this period due to a shortage of large-tusked males, this number of dead elephants has to be an underestimate, as it does not account for the calves and family members that would have perished after their mothers and family elders were shot and killed.

The vast majority of this ivory being taken from Africa was imported into Britain. In some years of Queen Victoria's reign, Britain was importing as much as 650 tonnes or more, and the figure averaged around 480 tonnes per year throughout her reign. Britain was using the ivory from around 30,000 dead elephants a year to make knife handles, piano keys, combs, and billiard balls. Big game hunting was all the rage and white elephant hunters would kill hundreds of elephants routinely. One man, the Scottish hunter Walter Dalrymple Bell, is reported to have personally killed over a thousand elephants during his safaris at the turn of the twentieth century.

Of course, it wasn't all wanton greed and entitlement. The Victorian-era administrators did have some understanding of sustainability, with various regulations imposed at differing times in different regions to try to protect and maintain what was visibly becoming a scarcer resource, usually centring around minimum tusk weights that could be harvested. But as we can see from the numbers, these regulations did little to halt the plundering continent-wide.

A far more significant step taken during the late nineteenth century was the proclaiming and gazetting of parks or reserves, aimed at protecting the natural habitat, flora, and fauna of certain landscapes. Late Victorian-era administrators realised that for there to be anything left for them to hunt (and trade), they would have to start preserving some animals.

In areas such as 'British East Africa' (Kenya, today), protection of wildlife went hand in hand with hunting – a conservation model that persists to this day in many parts of Africa. From 1896, game reserves were established in British East Africa to protect the land and animals, so that European colonialists would still have something to shoot, and the Game Department was set up specifically to protect these exclusive hunting grounds.

The establishment of reserves had a different origin in southern Africa. By the turn of the last century, excessive hunting had virtually eliminated elephants from South Africa, as we saw in the Addo region. In 1898, President Paul Kruger of the Transvaal Republic (which falls in the north-east of modern-day South Africa, and against whom Britain fought the Anglo–Boer wars) thus proclaimed a no-hunting zone in an area in the far north-east.

The elephant population in this area was almost wiped out, with the numbers being not more than double figures. But by 1926, this no-hunting area had grown considerably in size; it was proclaimed a national park and renamed Kruger. It opened to public vehicles in 1927. At 350 kilometres long by 60 kilometres wide, this vast area – some 20,000 square kilometres – is about the size of Wales.

Despite its successes, Kruger National Park has faced many attacks to its vision over its history, with hunters lobbying for access to the park; farmers (such as soldiers returning from the First World War) wanting agricultural land to work; gold, copper and coal prospectors wanting to mine the land; and – around the turn of the twentieth century – vets calling forcefully for the mass slaughter of all wild animals in the area to halt the spread of diseases carried by tsetse flies.

Livestock sleeping sickness spread by tsetse flies was a big deal. Wildlife species are carriers of the disease but generally remain unaffected by the symptoms, while horses and cattle are highly susceptible. At the time, some claimed it was the biggest brake to development in the South African region, limiting transport routes and agriculture, as livestock could not survive in many areas. The argument was that if wildlife was culled – thus remov-ing the carriers and food sources of the tsetse fly vectors – the flies would be eliminated, and the cattle could then move into new areas conveniently left vacant. But the warden of Kruger at the time strongly opposed this strategy, and managed to protect the animals in the park from the policy.

By 1935, Kruger's elephant population had increased to about 135 individuals, and stood at nearly 1,000 by 1958. Today, Kruger has one of the most robust elephant populations on the conti-nent, despite continued calls from wildlife managers and scien-tists in South Africa who vociferously warn of the potential damage of 'too many elephants', as we shall see in the following chapters.

One thing that the South African parks did have in common with those in east Africa, was the view about who owned the wildlife and who should have access to it. In both regions, parks and reserves served to 'protect' wild animals and their habitat from native populations. Africans were kept out of east African parks so that white people did not have to share their hunting grounds, and they were kept out of South African parks so white people did not have to share anything – remember, these parks were being established at the same time as apartheid policies were forming and snowballing.

This ownership of elephants by states and entities therefore signalled the beginning of the transition from the majority of hunting to fuel the ivory trade being legal, to illegal killing – i.e. poaching. But more significantly, these policies of ownership and exclusion have ramifications for conservation across the entire continent that still manifest today.

The actions of trophy hunting and trade in ivory hugely depleted elephant numbers up to the early twentieth century, but the very high population numbers initially across the continent meant the rate of decline was fairly low at first. Ivory prices and demand fell during the period of the two world wars, and there is some evidence that elephant populations across the continent began to recover and increase in number from around 1914 up to the late 1940s. But by the 1950s, demand for ivory increased again and the killing resumed, such that by the 1960s, the killing was at a level similar to the 1880s.

Our desire for ivory exploded even more in the 1970s, which is when the total elephant population began to fall dramatically. This coincided with the increasing wealth in the global north after the austerity of the post-war years, and a boom in the world human population – particularly across Asia, which in turn fuelled a new and expanding market for ivory in places like China and Vietnam.

Little scientific data on elephant numbers exists before the 1970s, and our population estimates up to this time are mostly based on calculations from ivory export information. Real, direct data on the sizes of elephant populations remained scarce until 1979, when Iain Douglas-Hamilton (the ultimate hero for all elephant biologists) undertook the first continent-wide survey of elephant populations. He pioneered the use of multiple aerial surveys to count elephants across their range, and his data from these counts combined with analyses of other kinds of survey gave an estimate of only 1.3 million elephants remaining across Africa in 1979. Worse still, these elephants existed in very fragmented populations – with few or no clear pathways connecting them.

As if the killing of elephants wasn't already out of control, it was made significantly worse by the introduction of automatic weapons in the 1970s. This coincided with a number of wars, civil disturbances and guerrilla campaigns that were occurring as a result of decolonisation and the chaos that ensued.

The value of arms imports to Africa increased tenfold, to billions of dollars, in the space of one decade up to 1980. As the last vestiges of colonialism were overthrown and dozens of African countries got their independence, there was a surge in violence across the whole continent. Much of this was due to arms races meant to defend or protect against adjacent countries that were increasing their own military power.

As the Ugandan military expanded under Idi Amin, neighbouring Tanzania and Kenya felt compelled to better arm their own forces. An increasingly militarised Libya egged on Sudan to expand its capabilities, and similar actions can be seen with Somalia and Ethiopia. Armies in east Africa grew from fewer than 140,000 personnel in 1971, to 440,000 in 1980.

Sometimes, these armed soldiers were themselves responsible for the poaching. In other cases, weapons would have been lost, stolen, traded, or forcibly removed from defeated soldiers, to end up in the hands of poachers. Whereas in the first half of the twentieth century, poaching was mainly done with spears, arrows and antique firearms, from the 1970s the AK-47 became the weapon of choice.

In Chad, for example, the elephant population declined dramatically (falling by 80 per cent) within a few months of the outbreak of war in 1979. The long civil war in Mozambique, lasting from 1977 until 1992, had a devastating effect on elephants and other large mammals. Army, militia and the 'resistance' all poached elephants in the areas under their control, for both ivory and meat. In Gorongosa National Park alone, elephants declined from several thousand individuals before the conflict, to about 300 by 1992.

I've seen it myself in South Sudan, a country that used to have a vast and diverse range of wildlife, now almost wiped out because of poaching for trade and bushmeat during their civil war. The perpetrators are almost always militiamen armed with AK-47s and I've had to witness many occasions when youths thought it would be fun to fire a volley at a herd of antelopes or giraffe to see if they could get some extra meat for their dinner.

Similar stories can be found in many other African nations, and the effects of such conflicts can be so severe for wildlife that some conservationists have called recently for explicit safeguards for biodiversity during armed conflicts, with environmental damage cemented as a war crime under the Geneva

Convention. They want international legal instruments to protect natural resources from armed conflicts, to regulate arms transfers to poachers, and to hold militia and military personnel accountable for the environmental damage they cause, be that destruction of megafauna, their habitats, or poisoning of natural water sources.

By 1987, elephants had been all but eliminated from areas of Sudan, Chad, Central African Republic, and Zaire (now DRC), and had been greatly reduced – and were still declining – across east, southern and west Africa. Only the populations in Malawi, Botswana, Zimbabwe and South Africa seemed robust at the time. Two years later, the continental elephant population had fallen further, to around 650,000 individuals. The number of elephants had halved in one decade, and the ivory yield was three times higher than the maximum sustainable level. The speed of the decline was so steep, because there were many fewer elephants to begin with than in earlier centuries.

By the late 1980s – when I was a young child – the situation was so severe that it was predicted that elephants could be extinct across Africa by the new millennium. I can distinctly remember campaigns such as 'EleFriends', and the 'Babar says Help Save the Elephants' window stickers, from the awareness campaign championed by Iain Douglas-Hamilton.

Luckily, the campaigns worked, and in 1989 an international ban on ivory trading was imposed. The ban had the desired effect, with demand for ivory falling almost instantly. The industrial-level killing stopped, and elephant numbers began to recover again. But of course, the ivory story does not end there. For reasons we explore more in the next chapters, demand for ivory began to increase again by the late 2000s, and intense poaching resumed. Around 30,000 elephants were being lost each year between 2010 and 2014, reaching a peak in 2011 with 40,000 elephants killed. And the killing continues today.

Taking up Iain Douglas-Hamilton's mantle, Mike Chase, the founder of Elephants Without Borders (where I visited the orphanage in Botswana), undertook the Great Elephant Census (GEC) between 2014 and 2016. It is the largest count of elephants to date, with aerial censuses occurring over eighteen savannah elephant range countries. Mike aimed to count around 90 per cent of the elephants in the eighteen countries, and compare the numbers of live (and dead) elephants counted to what we know about historic population sizes.

Mike uses the latest in GPS technology to track elephants as they migrate. In the summer of 2019, I joined him in an aerial recce mission to dart and collar a bull elephant in the Okavango Delta, as part of his ongoing research. It was a thrilling experience. Mike drove in his trusty old Toyota Land Cruiser through the bush and instructed me and his friend, the veterinary doctor Larry Patterson, to fly in a helicopter around an area near to Seronga.

Almost immediately after take-off, we spotted two adult bull elephants in their late twenties in a clearing; they were slowly walking and grazing together as we swept around to encircle them. Larry spoke to Mike on the radio, telling him where to drive. Once Mike had spotted the elephants from the ground, he drove as close as he could get safely and chose which one to target. Larry loaded his rifle with a dart filled with carfentanil tranquiliser drugs and we banked around to try to get a shot. The helicopter pilot, used to these sorts of missions, joked as we swung over to the right – the side door was fully open, and I was glad we were strapped in.

'Closer, steady now,' shouted Larry over the mic. We hovered at around fifty feet in the air as the older elephant stormed off into the treeline, while the younger one ran around in circles. That's exactly what we wanted, to separate the pair.

'He's scared, but he'll wait for his mate in the bush,' said Larry. 'If there's danger, they will leave their mates behind for a bit, but they'll be reunited soon.' And then, as soon as Mike was in position, he gave the command to shoot.

Larry took aim from the wobbling chopper and fired a shot. 'Got him,' he smiled. The helicopter soared upwards and began to circle again from a safer height. We waited for the drug to work. 'It usually takes fifteen minutes.' So we circled around as the bull gradually slowed down, and after a while he came to a complete standstill.

'He's a tough bugger,' said Larry, after twenty-five minutes. The elephant was still standing upright, although he was clearly in a trance. 'We'll need to give him another shot. Let's land,' he said. So the pilot came down, landing the helicopter fifty yards away in a clearing near to where Mike was parked, calmly watching.

'Yeah, he's a tough one, that's for sure,' said Mike, as we all walked over to the weird figure of a drugged elephant standing completely still. 'Is it safe?' I asked cautiously, as the beast towered above us. It felt very surreal to be in the shadow of this enormous creature that appeared to be suspended in time.

'He's out of it,' said Larry, 'but better safe than sorry.' With that he walked up to the elephant and jabbed another dart in his backside to make sure we had a clear thirty-minute window to do the job. The elephant wobbled a bit, then crumpled under his own weight, luckily falling onto his side with a loud thud that shook the ground. If he had fallen forward onto his chest, we would have all needed to push the big male over so as to prevent the lungs being crushed, but fortunately that wasn't necessary.

As soon as the elephant was still, we all set to work. Firstly, Larry pulled an ear over the elephant's face to prevent the sun shining directly onto his face and flies from landing in the sleeping giant's eyes. Then Larry began measuring the elephant's dimensions – the tusk length and estimated weight, the skin condition and overall shape that he was in. He was remarkably fit and healthy, especially given the recent drought that Botswana had been undergoing. 'That means he's either been on the move a lot looking for new food. Or else he's stayed in the delta where it's green,' said Mike.

As Larry did his thing, Mike and I set about putting on a collar, which involved lots of pulling and shoving until the weight was absolutely right. 'This thing weighs four kilos,' said Mike, holding the big lump of lead that acts as a counterweight to the tracking device, which sits on the back of the neck. I worked as fast as I could to screw the bolts in place, and using some shears, chopped off the excess rubber. I dreaded to think what would happen if the bull woke up prematurely. 'Don't worry,' said Larry. 'We've got another eight minutes. Loads of time.' It didn't reassure me!

But with the collar fitted, Larry injected a dose of the tranquiliser antidote to make sure the anaesthetic would wear off in good time, and we retreated to a safe distance as the animal rolled up and gradually got to his feet. No doubt with a beastly hangover, he shook his head and realised that he now had a new necklace, much to his displeasure. The bull started tugging on it and tried to pull it off.

'They wonder what it is, but after a very short time they get used to it. It'll stay on for three or four years, and then either drop off naturally, or we will come and remove it.'

As the bull ambled off to find his friend, I appreciated how much effort went in to gathering all this data. Mike and Larry had done this literally hundreds of times. But that was only the beginning. Then the scientists had to track and follow each elephant and find out exactly what they were up to.

His data, sadly, does not paint a pretty picture.

In the last thirty years, the elephant population has fallen again by another quarter, to reach its lowest known figure of around 415,000 elephants across the whole continent. Currently, populations are shrinking continent-wide at about 8 per cent per year, and this is still primarily due to poaching.

Here, we must insert a caveat: Africa is not monolithic. It is a hugely diverse continent comprising of fifty-four (currently recognised) individual countries, that each have their own geography, demography, and ecology. Each country has its own programmes, plans, and priorities for conservation and development, and we should not fall into the trap of talking about problems with elephants as though they are universal. The biology of elephants may be much the same across the continent, but the situation of each country – if not each administrative or ecological area within a country – is unique, and therefore the difficulties facing elephant populations vary greatly across their range.

That being said, there are patterns or trends in what has been happening that are common across several or numerous countries. Only bear in mind that what has happened in one population, country, or region may not represent the story for all elephants within the continent.

In west Africa, elephants are now extinct in Guinea, Guinea Bissau, Sierra Leone, and Togo, and the only countries in the region to have populations that number more than a couple of hundred elephants are Benin (at nearly 3,000) and the contiguous population of Burkina Faso (6,850), and Ghana (at just under 1,000). The continent's westernmost population of elephants – in Senegal – is estimated to be ... one. It was seen in 2013.

In east Africa, populations in Kenya, Uganda and Malawi are showing fairly positive trends in recent years, but certain populations in Mozambique and Tanzania have experienced very large declines, with some of the greatest poaching rates in the continent between 2008 and 2016. In southern Africa, however, stable or increasing elephant populations were recorded in Botswana, South Africa and Zimbabwe. Some of these increases are due to migration, as we shall see in subsequent chapters, with elephants moving into these relatively safe areas from other locations that have been heavily hit by poaching, such as Angola.

Forest elephant numbers are almost impossible to ascertain, given the dense vegetation cover in the areas in which they live, making aerial surveys impossible, but between 2002 and 2011, elephants in the central forest block (from Gabon and Republic of Congo right across DRC) were counted on numerous walking surveys.

These surveys estimate that there are probably only 30,000 left and show that their population declined 62 per cent between 2002 and 2011, and 30 per cent of geographic range was lost in the same time period to agricultural development and land conversion. In less than fifty years, forest elephant numbers are thought to have fallen by more than 80 per cent. That is, less than one-fifth of the 1960s population of forest elephants now remains.

Forest elephants now inhabit less than 25 per cent of their potential range, and likely number only 10 per cent of their potential population size. Poaching, increasing human populations, rapidly expanding infrastructure, poor governance and law enforcement, all contribute to these problems. The scenario typically is that logging companies move into an area, building roads that open the dense forest to further encroachment and allowing easy access routes for poachers. These roads also permit illegal logging: it is not only animal parts that are trafficked – a huge illegal market also exists for African hardwood trees; and as the forest is chopped down, so are the last remaining habitats of the vanishing forest elephant.

I was once lucky enough to visit the DRC, travelling through some of the last volcanic upland forests left in Africa. It was an undoubtedly beautiful environment, and yet it was tragic to see so little left. Farming and ever-growing villages were encroaching all around the protected areas, and even inside the national parks, local people and gangs of poachers were relentlessly placing traps and hunting the last vestiges of wildlife, not to mention chopping down ancient trees and burning whole segments to make way for crops.

If you look at Virunga National Park on Google Maps, you'll see how fragile an environment it really is. Straight lines show the scars of human invasion, and the ever-decreasing forest left for these little elephants in which to roam. All the surrounding countryside in DRC is cultivated, and everything on the other side of the border in Rwanda appears like a dirty patchwork quilt of maize and bean fields.

It's almost impossible to know how many elephants are left there, as the rangers are few in number, inexperienced in elephant behaviour, and are constantly threatened by armed gangs and murderous militias, who try to kill them when they go out on patrol. In fact, hundreds of rangers have been killed in the last few years in the line of duty. An aerial census would prove fruitless, as the forest canopy is too thick and whatever elephants there are remain invisible.

I guessed there might be room for four or five herds in this forest, probably totalling no more than a couple of hundred, but it may have been much fewer. At least there they could range over the mountains into Uganda and Rwanda, and hide for a while in the jungle valleys away from the greed of humans, but ultimately they were hemmed in on all sides, and unless the forest is protected, it won't be long before there are no forest elephants left to count.

Savannah elephants are much easier to spot. The Great Elephant Census (GEC) and Mike Chase's subsequent analysis show that elephant numbers increased a little between 1995 and 2007, but they have been declining at a continental level ever since then, with especially rapid declines between 2010 and 2014. The majority of individual elephant range countries follow this same pattern, with the number of deaths far outstripping the potential number of births. The GEC also showed that most elephants now live inside parks or reserves – with 84 per cent of elephants counted being in protected areas. Only in Mali and Angola were more elephants counted outside protected areas than inside.

Given the high rates of poaching observed, the only conclusion is that many reserves are failing to protect their elephant populations adequately. In fact, if populations continue to decline at this rate, they will halve every nine years.

Perhaps what is even more astounding than all these figures, showing how quickly we have destroyed elephant populations, is the fact that data repeatedly shows that savannah elephant numbers can and do recover, given the right opportunities. We've seen it in Addo and Kruger; it has happened in Etosha, Namibia; and we have some reason to hope it is happening in other populations.

I briefly visited the Tarangire National Park, in north-east Tanzania, in the autumn of 2010 while driving down to Malawi. It's a beautiful place filled with fat baobab trees scarred by centuries of wear by elephants, who enjoy coming to scratch off the bark with their tusks.

The region was affected by heavy poaching outside the national park boundaries in the 1970s and 1980s, especially on the main road that dissects the protected areas, and many elephants moved into the relative safety of the park itself, so the population increased from 440 in 1960 to about 2,300 in the year 2000. The poaching of older bulls meant this population had a particularly high number of young females.

Between 1993 and 2005, the elephant population was increasing in number at 7.1 per cent per year – the maximal rate of population increase for elephants. This high rate was achieved after intense poaching, because the females in the population had very short gaps between calves, they were young when they had their first calf, and death rates were also low. Females were literally pumping out calves as fast as they could, helped along by some years with very good rainfall, which meant lots of grass was readily available, giving them plenty of energy to reproduce.

When I was there, the Tarangire population was around 3,000 elephants, so this population has managed to continue increasing despite the recent poaching surge that has devastated other populations in the region.

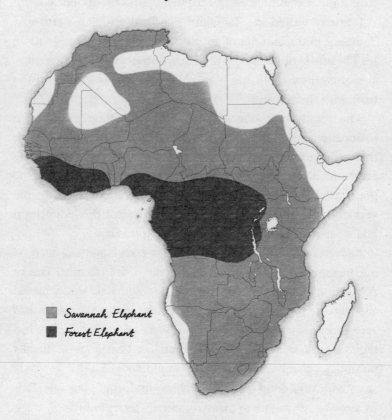

Historical Range of Elephants c. 1800

Savannah Elephant
Forest Elephant

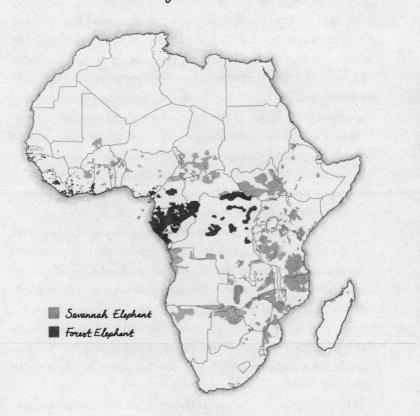

Present Range of Elephants 2019

Savannah Elephant
Forest Elephant

But sadly, the Tarangire example is not universal. In Mikumi National Park, also in Tanzania, 75 per cent of the population was lost in heavy poaching in the years up to 1989. By 2004, a third of all adult females were still not reproducing and group sizes remained very small. It has been suggested that the key difference between Tarangire's recovery and the lack thereof in Mikumi, is that by 1989, only 14 per cent of family groups in Mikumi had a matriarch that was over thirty years old (compared to nearly half with an older leader in Tarangire) showing how important the female leaders are to the species survival.

When you mess with family structure, it can have untold consequences, and it goes to show that numbers alone do not tell the whole story. Let's imagine some anthropologists find a previously undiscovered island inhabited by a tribe of sixty-year-olds with their children and grandchildren.

We might assume, because there are old people on the island, that the young have had decent role models, that they are well informed, skilled in bushcraft, sociable and well adjusted. The anthropologists can see that the young don't eat certain foods, because the elders have passed on some explicit knowledge about what is and is not good to eat, and they harvest other foods in certain ways or at certain times of the year for the same reasons. The anthropologists also see that they act in unusual ways that don't seem to make sense.

Then after a few years of studying them, the anthropologists have noticed that the harvest is not very good, in fact it's damaging, unhealthy and unsustainable, and that many potential food sources are ignored for no good reason. Not only that, but the social dynamics between the island inhabitants are a bit off; the children are rebellious, couples argue, the people are prone to disease and the birth rate is low.

Then the anthropologists discover that those sixty-year-olds grew up on that island without any elders themselves; perhaps they were shipwrecked as children or some disease tragically killed all but one or two of the adults in the population fifty-five years earlier. So, as a result, today's adults did not acquire any social traditions or knowledge about what to eat, when and how, but instead made it all up themselves as they went along.

We would not be at all surprised if children who grew up without elders, without any acquired traditions or cultural knowledge, resulted in a dysfunctional, or at least sub-optimal society. Yet we almost constantly underestimate the effect this could be having on elephant populations across Africa. We have been altering elephant habitat, ecology, population genetics, demography, and thus their social structure, for hundreds of years. Not merely altering a little bit, but really screwing it up.

Many of the elephant populations we see today are oddly structured, inbred, or have recovered from very small population numbers, which means they grew up with no guidance and very little hereditary knowledge. The lists of numbers and declines that have appeared in this chapter can be a bit theoretical and too large to fully comprehend, and it is easy to lose sight of an important underlying point here.

So few elephant populations are untouched, we don't have much idea what is normal any more – but more critically, perhaps, neither do the elephants that are left. We have forced so many of today's elephants into a situation where they simply have to make it up as they go along, like the survivors on our pretend island, and we can't yet quantify or understand the implications that this could have.

Rates of killing elephants peaked during the 1970s and 1980s, but the twentieth- and twenty-first-century killing has caused much greater population reductions, because the elephants were being taken from an already smaller population to begin with, made even worse by the fact that the populations that do remain are potentially scarred socially and emotionally beyond repair.

The figures show clearly that the colonial period and the slaughter in the early twentieth century was responsible for a steady and substantial decline in elephant numbers across the African continent, but the decimation has been even faster since the present century began. It's now time to look at the main reasons for this decline, starting with the one that has historically been the main reason that humans have killed elephants for millennia: our insatiable desire for their teeth.

8

Poaching, Ivory and Trade

My grandfather, the same one who used to take me on a blind-folded journey into the 'elephant graveyard' as a child, was also an unwitting contributor to the deaths of unknown numbers of elephants himself.

He was the proud owner of an old 1920 knock-off Steinway upright piano that, of course, had ivory keys. He also had a number of horn- and ivory-hilted knives and forks, part of his prized collection. He wasn't alone. Until the 1950s, before the widespread use of plastic, most keyboards had ivory keys, and ivory was a fashionable accessory for jewellery and decorative art, highly prized by all classes around the world. It was this trade that ultimately led to the rapid decimation of elephant numbers and it would be impossible to fully understand the plight of elephants today without considering the importance of that trade, both legal and illegal, over the course of the last century.

As we have already seen, the ivory trade has reduced the total African elephant population from many millions to a little over 400,000 individuals – less than the human population of a mid-sized city like Bristol. Whilst the mass killing of elephants has mostly been illegal in the last half century, the international trade in ivory was only outlawed in 1989.

Since then, two 'one-off' (yes, a total oxymoron) international sales of ivory have been permitted, in 1999 and 2008, between a few countries, as we shall see below. And domestic trade within countries has been largely legal. Domestic ivory trading bans have only been enacted very recently in a handful of countries, including the UK, which introduced domestic ivory trade legislation in 2018.

But if international trade in ivory is illegal, and domestic trade is increasingly being curbed, why are we currently living through another elephant poaching crisis? Why did poaching surge again from 2008? Why did 40,000 elephants die in 2011? And why have more than 600,000 kg of ivory tusks been seized in illegal shipments to Asia, Europe and the US in the years since the 1989 ban?*

The illegal wildlife trade – of which ivory is a part – is one of the principal threats to wildlife in Africa. The trade is currently the fourth or fifth largest international criminal industry after drugs, counterfeiting and human trafficking, and is worth as much as US$23 billion per year according to the UN. It's around the same as, if not more lucrative than, the illegal arms trade. Not only does the trade threaten the existence of species, but also international security, national sovereignties and impoverished rural communities.

Understanding – and controlling – the ivory trade is no simple feat. We need to consider the trade from all angles; from the poachers and the opposing individuals and agencies charged with protecting elephants on the ground, to the people coordinating

---

* At an average of 12 tonnes per year, these seizures are obviously much smaller than the weights of ivory that were being legally exported from Africa in Victorian times. But remember that the amount that is seized by customs and law-enforcement officials represents only a portion of the total ivory that is being smuggled out of Africa now, and that as tusk size has decreased, the number of elephants that have died to make up one tonne of ivory is now likely much higher than it was in the nineteenth century.

the mass movement of illegal produce, and the consumers who fuel the trade by buying ivory. But of course, we also need to consider the trade from the elephants' perspective as well.

Tusks – ivory – are simply teeth. Granted, different from our teeth, but they're still teeth at the end of the day. Tusks appear from around two years of age and can grow at a rate of more than 10 cm per year in male savannah elephants, slightly less in females.

Some of the largest tusks ever known are those of Ahmed, a magnificent male who lived around Mount Marsabit in Kenya, in the mid-twentieth century. Ahmed's tusks each weighed 67 kg and measured almost three metres long! Some brilliant legends surround Ahmed, my favourite being that his tusks were so large that he had to walk uphill backwards. Ahmed was such an important symbol that Kenya's first president, Jomo Kenyatta, conferred presidential protection on him, with five armed rangers providing constant security.

The tactic worked and Ahmed, despite being in his prime during the poaching resurgence of the 1960s and early 1970s, died a natural death in 1974, aged around sixty-five. During his autopsy, bullets from antique British Army rifles were dug out of his body, suggesting his life had not been entirely straightforward and his special protection was warranted. Few such 'big tuskers' remain, certainly no more than an isolated handful in South Africa and Kenya.

One now-famous big tusker of our lifetime with a less happy story is Satao. Although not quite as big as Ahmed's, his tusks were still immense – 50 kg each, and so long they swept the ground in front as he walked. He was perhaps the largest tusker left in Africa in the twenty-first century. Amazingly, Satao managed to live in obscurity in the vast wilderness that is Tsavo for most of his fifty years, unknown even to long-term researchers and conservationists in the area. But his home range bordered the edge of Tsavo National Park, in a region that was increasingly being overrun with poachers.

In 2013, Mike Chase, of Elephants Without Borders, spotted Satao from the air, when he was surveying the area as part of the Great Elephant Consensus. Mike was concerned that Satao might be injured, having noted that he was moving in a strange way. Sure enough, veterinary teams that were dispatched found that Satao had been hit by a poacher's arrow, but the wound was not deep or life-threatening. They decided not to undertake the darting procedure – made especially risky given Satao's size – and Satao did recover quickly on his own. The bad news was that this meant the poachers in the area knew about him and were unlikely to give up on such a prize.

In May 2014, the arrival of rains – and with them, lush vegetation – drew the normally cautious and reclusive Satao to the edge of the park in close proximity to local poachers. Here, he was hit for a second time with a poisoned arrow, and this time he was not so lucky. The arrow penetrated deep into his body, and the poison rapidly took effect; he collapsed and died soon afterwards. He was found a few days later by conservationists, with his face hacked off and tusks nowhere to be seen.

His death would have been a significant bounty for the poachers who escaped with his tusks – earning them a couple of thousand dollars on the black market. Yes, you read that right: Satao was killed for probably less than the average British family spends on holidays per year. However, there's an impressive mark-up in price along the whole ivory chain, given that end-consumers in China were paying about $2,100 *per kilo* in 2014, which would have made Satao's tusks finally worth over $200,000.

Interestingly, a filmmaker who spent four years living in Tsavo whilst making the beautiful documentary film *The Elephant Queen* was the first to record and photograph Satao (although not publicly – the filmmaker kept all imagery and information about Satao under wraps in order to protect him before his death). He was convinced that Satao tried to hide his massive tusks as he walked. Satao travelled by zigzagging between bushes, and shoving his head and tusks into each bush as he reached them. There he'd wait for a few minutes, before continuing on to the next bush. Did Satao really understand what a liability he was carrying?

I guess we'll never know, but one thing is clear, Satao's death – perhaps more than that of any other elephant in Kenya – galvanised the Kenyan people into taking action against the loss of their iconic wildlife, piling pressure on President Uhuru Kenyatta, who since then has been one of the most vocal leaders in anti-poaching campaigns. And since Satao's death, Kenya's elephant population has been better protected, with more rangers recruited to the Kenya Wildlife Service, more funding for KWS anti-poaching activities, and better coordination between state and civil-society conservation and law enforcement organisations. The result is that elephant poaching in Kenya has been decreasing recently, and the elephant population is finally stabilising.

The scarcity of tusks like Satao's nowadays illustrates an intrinsic problem with poaching: the intense selective removal of large tusks means that fewer and fewer animals with large tusks are left. But even more than this, rates of tusklessness are increasing – that is, we are seeing more and more savannah elephants with no tusks at all. It is Darwinism at its most ruthless.

Under normal circumstances, we would expect about 2–4 per cent of females to be tuskless – it happens, but it is comparatively rare. But in Gorongosa in Mozambique – a population that has been heavily poached for a prolonged period – more than 50 per cent of older adult females that survived the civil war are tuskless, and more than 30 per cent of females born since the war are also tuskless. In Ruaha National Park, Tanzania, the figure is 21 per cent, and higher in older age classes. And in Addo, whose current population, you will remember, was founded from a bottleneck of only eleven elephants, 98 per cent of the females were tuskless in the early 2000s.

Tusks are used for obtaining food, such as tree bark, or digging for water, as well as in aggressive encounters between males. The impact of having more elephants born without these tools is not yet known. Perhaps elephants will have to range over larger areas to find sufficient substitute foods to replace what they could have accessed if they had tusks, or maybe they are finding innovative solutions to access food in other ways. It could also have an impact on the environment; failing to dig for water could mean that other animals who rely on the dug pools go thirsty, or lizards that prefer to live in trees that get tusked and de-barked by elephants could find themselves without enough suitable homes. The implications could be far-reaching and enormous in the long run.

But poaching affects more than rates of tusklessness. As we saw with the Mikumi population in Tanzania, if too many older individuals – especially the wisest matriarchs – are lost, breeding success and population growth can fall off dramatically. Socially, the young survivors of poaching are rather clueless, floundering around when they do not have elders to guide them. Orphaned calves who have survived poaching raids tend to form less-structured groups, often with other, unrelated individuals – as I found in the Botswana orphanage with Panda, Molelo and Tulli. The remnants of families who have lost their elders cluster around the oldest individuals that remain, while stress levels are also much higher in groups that do not have an experienced matriarch.

The groups formed by orphans can be large, but they are fairly loose – more of an aggregation than a unit. Like living in a crowd of neighbours, not a family. Moreover, those orphans are so socially disadvantaged – receiving more aggression and less affection when growing up without their mother – that they are less likely to survive long-term, or to be successful mothers if they do survive.

It can take an elephant population many generations to overcome the social effects of poaching, yet these real-life consequences are often lost in the seemingly endless debates about the politics and economics of the ivory trade. Because, of course, as soon as we use the word trade, economics becomes important. It's all about supply and demand.

East Asia, particularly China, is well known for being the epicentre of ivory demand nowadays (although remember it was only a century ago that Britain held that inauspicious title). Ivory carving has a long tradition in China and Japan, and great cultural value is placed on ivory works of art. In such countries, owning ivory is a status symbol, much like owning diamonds or fast cars, and so it is not surprising that as Asian economies grew and personal wealth increased, so did the demand for ivory. And as demand increased, so did rates of poaching, to maintain supply.

Asian elephants are even more endangered than African elephants, with only 40,000 left. As elephant numbers in Asia dwindled, it became more and more difficult for poachers to kill them, and so they turned to Africa for their supplies. Demand reduction is therefore key to reducing poaching in Africa. To borrow a phrase used by many anti-poaching charities: 'When the buying stops, the killing can too.'

There is some good news that this is already beginning to happen. Economic growth in China has slowed since about 2012 – the Chinese economy is still growing, and much faster than the US, UK, or other European economies – but at a slower rate than between 2008 and 2011.

Coinciding with a slowing Chinese economy, poaching rates have fallen slightly. They are still too high, but thankfully they are lower than the peak rate recently observed in 2011. The economic slowdown in China may well mean that people are less keen to buy luxury status symbols, so demand is falling. But this is a very precarious state, and we can't assume it will last too long. As soon as the Chinese economy picks up, so too might demand, and therefore the killing. Elephants can't afford for us to rely simply on global economic slowdowns to protect them.

Anti-poaching strategies vary across Africa. These range from direct action in the form of on-the-ground enforcement; wildlife protection agencies, ranger services and military-style protection units. Even within these organisations there is massive diversity in tactics, operational procedures and goals. Some are armed, others are not. Some have a shoot-to-kill policy, whereas others rely on a mere deterrent. Some prefer uniformed guards, and others implement undercover policing.

Several countries in Africa are trialling better contraband detection systems at transit points, in airports and border crossings. There are also education programmes in place, trade bans, and more and more ways of addressing the root causes of poaching by alleviating poverty in supply countries.

Within 'supply countries' – elephant range states – poaching may be perpetrated on different levels. At the top end there are the organised criminal networks, funded and heavily armed by international syndicates, often linked to corrupt government officials and the regional mafia. These gangs commonly employ individuals or small local groups, who sell the ivory they take on local black markets. These petty criminals are often nothing more than neighbourhood criminals armed with illegal rifles.

Then at the bottom end are the members of tribes, simply keen to escape from poverty, and sometimes armed with only a bow and poisoned arrow or a bag of cyanide. And yet, it is frequently the ones at the bottom, usually with the best justification, that do the most damage, because of the scale of the problem. In countries like Zimbabwe, there have been massive spates of killing that use this prosaic yet deadly method. Poachers bait oranges or even salt licks with cyanide and leave it out in an area where they know that elephants will pass.

The unwitting elephants can't resist such a treat, and the whole group may eat the poisoned food; entire families or groups then die very painful deaths, even youngsters with no or very small tusks. The killing is indiscriminate and on a mass scale. And to make matters worse, scavengers such as lions, jackals, and vultures that come to feed on the elephant carcasses will then be poisoned too from eating cyanide-laced elephant flesh. Vultures in particular, some species of which are already highly endangered, have been dying in large numbers following such poaching events.

Some poaching is less ad hoc, and even darker – being linked to rebel militia groups. Heavily armed militia from South Sudan, Chad, or northern Uganda have perpetrated many mass killings in central Africa.

In 2012, several hundred elephants were killed in a few months in Cameroon, by well-armed poachers riding on horseback – thought to be Sudanese or Chadian militia. The LRA (Lord's Resistance Army), led by Joseph Kony and infamous for its violent, forced recruitment of child soldiers, is known to have perpetrated much of the poaching in Garamba and Dzanga between 2011 and 2014. Al-Shabaab militia have also been implicated in some of the poaching occurring in east and central Africa, and the resultant funds have been linked to supporting terrorist organisations such as al-Qaeda and ISIS.

But whether they are acting as part of organised militia groups, or at a more 'subsistence level', the increasingly sophisticated weapons being used are extremely bad news for the rangers and staff charged with securing protected areas. Rangers are typically individuals who are interested in wildlife, perhaps graduates of biology, ecology, or tourism courses. Even though they often wear camouflage gear, and many even carry weapons (originally meant to protect them from dangerous animals), they are not usually military personnel. Yet despite their lack of training and combat experience, they have progressively become engaged in an armed conflict where they are expected to fight and shoot at other human beings.

Many reserves now employ special police or army personnel to work alongside and train rangers, as the arms race between poacher and wildlife protection officers has escalated. There is a growing corps of international volunteers, often former military officers and special forces operators from Western armies, who mentor these African wildlife agencies. It has even become an official mission for the British army, who now send teams of regular soldiers to assist in anti-poaching efforts in countries like Gabon, Malawi and Zambia.

It's a positive move when it comes to implementing immediate action and helping the operational effectiveness of what might otherwise be meaningless academic chatter or political hot air, but it also shows how desperate the situation has become. Many countries have even instigated 'shoot-to-kill' policies, which permit or even encourage rangers to take the life of poachers if they are caught inside protected areas. And of course, it endangers more and more people involved in conservation in the process.

The Liwonde National Park in central Malawi is a large wilderness area filled with thick acacia forests, large rocky hills and lots and lots of elephants. Like many British soldiers, it's an area I know well, having served there myself some fifteen years ago. The British army have used the remote park for a long time, as it's a perfect place to train in hard, hot and dangerous conditions before being sent off to war.

Sadly, the elephant herds inside the park have recently become easy prey for cross-border poachers from Zambia. For a while there was little to be done, as the criminals sneaked through the bush and stole off with a few tusks undetected. But when these gangs of poachers started arming themselves with better weapons, and the scale of the killing grew, the Malawian government requested the help of the British army to help stop the incursions. Soldiers found themselves going from a routine training exercise to a live operation.

Matthew Talbot, a twenty-two-year-old soldier in the Coldstream Guards, was one of those adventurous souls who was keen to gain as much experience as possible, and volunteered to help in the process. On 5 May 2019, he was on patrol through the bush. Having been trained in all the skills necessary for survival, he was keeping his eyes peeled for the deadly poachers that were on the loose. In a cruel twist of fate, that day he was tragically killed, not, as it happened, by a poacher, but by an elephant. The details are not as important, however, as the fact that he gave his life to help protect this species, as so many others have done before.

It is difficult to obtain accurate figures on how many rangers and poachers are being killed or injured in such conflicts across the continent, but one estimate suggests at least two rangers are dying *per week* in the line of duty. Not that it undermines any sense of the tragedy, but the number of police officers killed on duty in the UK averages fewer than two a year since 2000.

There are other human consequences of this militarisation, too. There have been a number of worrying reports that some already marginalised communities – such as the Baka 'pygmies' of DRC – have been victimised at the hands of government officials, receiving verbal and even physical abuse for merely being *suspected* of harbouring or protecting poachers. The psychological costs of these wildlife wars – on rangers, police, armed forces and local tribal communities – should not be underestimated or ignored.

It may be easy to disparage and detest the individuals who pull the trigger or lay the poison to kill an elephant, but many of them are simply seeking a decent income in a difficult world. Poverty is the key driver of poaching at the local level. The real bad guys, as with so many illegal activities, are the kingpins leading the organised crime syndicates that buy up, smuggle out, and sell on the ivory in bulk. And it really must be organised; the volumes of ivory being exported in blocks, passing through various customs points in regular transit routes, indicate some very sophisticated planning indeed.

Not all ivory reaches the consumer countries in this way – plenty of people are arrested at airports across Africa each year with a few small tusks or pieces of carved ivory bought on the black market shoved down their trousers, or hidden in a suitcase. But ivory has increasingly been seized in large shipments.

Reports from CITES (the Convention on International Trade of Endangered Species) – the global body tasked with tracking and regulating the international ivory trade, among many other things – show that whilst poaching rates have been in decline since 2011, seizures of illegal ivory are hitting record highs: 2013 and 2015 saw the greatest volume of illegal ivory seized since 1989. The countries of origin or elephant populations that the seized ivory came from can now be determined through the use of advanced DNA analysis techniques. As expected from poaching rates, much of it analysed so far comes from the Republic of Congo, Cameroon, Tanzania and Central African Republic.

This increase in seizures of large-scale ivory shipments could purely be a result of better detection and enforcement – i.e. more ivory is being found because customs officials are getting better at finding and reporting it. However, it could also indicate that the crime syndicates responsible for much of the illegal flow are smuggling more ivory, possibly because they are getting cocky, or perhaps getting twitchy, and they want to get it to the destination countries as soon as possible.

Syndicates obviously want to sell ivory for the highest prices possible, and the rarity of a product is a tried-and-tested way to increase its price. In essence, ivory syndicates have been speculating on elephants becoming extinct. They want their finite product to become even scarcer, so that the prices of their stockpiles become wildly high. But now these crime syndicates may be involved in a mass sell-off, recognising that as enforcement strategies improve, their risks of getting caught are more likely, following some high-profile arrests made recently in Kenya and Tanzania.

The syndicates may also be realising that the world is not going to allow the extinction of elephants that they were banking on, so the ivory they have stockpiled is never going to be worth the crazy sums they were hoping for – and as demand falls, its reserve value is decreasing all the time. So maybe they are thinking it is better to sell it now, getting it off their hands whilst there is still some demand.

Decreasing demand is almost certainly related to economic downturns, but I think it would be wrong to disregard the effects of public demand reduction and awareness campaigns across consumer countries. The global outcry about ivory poaching has been considerable and effective.

For example, the remarkably tall 7ft 6in Chinese basketball star, Yao Ming, teamed up with the charities Save the Elephants and WildAid to produce ad campaigns in China explaining what the demand for ivory was doing to elephants in Africa. The charity Tusk Trust, whose patron is HRH the Duke of Cambridge, has fought tirelessly for years to lobby the Chinese government to stop the trade. Before such campaigns, 70 per cent of Chinese people surveyed did not know that elephants had to die to relinquish their tusks for the ivory trade. This number now seems to have fallen substantially. And as more Chinese people realise that their desire for ivory is killing elephants, fewer want to own ivory pieces.

Bowing to growing public awareness and pressure, many nations have now banned their domestic ivory trade, or are in the process of doing so. Domestic trade in ivory has long been known to form an effective mask for poached and illegally smuggled ivory. In 2016, President Barack Obama instigated a near total ban on ivory sales within the US. It is still legal to own ivory – heirlooms and antiques can be kept and passed on – but sales of any ivory that is not antique (defined in the US as more than 100 years old) are prohibited.

Importantly, China then followed suit, announcing its own domestic ivory ban that was enacted in 2017. Sixty-seven state-sanctioned ivory-carving workshops and retail shops were closed in March 2017, and a further 105 were closed by the end of 2017. As soon as the intention to enact a ban in China was announced, ivory prices began to fall – from the peak of $2,100 per kilo of unworked, raw ivory in 2014, to $730 per kilo in 2017. And so far, the fall seems to be continuing.

The UK also drew up legislation to ban domestic sales of ivory in 2018, which should be enacted into law by the end of 2019, and the EU is increasingly being lobbied to consider similar bans. Prince William even called for all the ivory in Buckingham Palace to be destroyed as a show of his commitment to the cause.

It is not all good news, though. About one in five Chinese ivory buyers are considered 'diehard', having stated that the domestic ban will not stop them from buying ivory. At least some of the market in China now appears to be moving (illegally) to Cambodia and Laos – particularly the Golden Triangle, an infamously lawless opium-growing hotspot that spreads across neighbouring Thailand and Myanmar (Burma). And Japan, the other major ivory-consumer country, remains determined to keep its domestic market open.

International trade in ivory has been criminalised since 1989, so almost all new ivory entering retail markets outside of elephant range states is illegal. Almost. Of course, hunters can import trophy tusks, requiring explicit CITES permission to do so, but as they are personal trophies, they do not usually end up being traded in domestic markets. However, substantial quantities of ivory have *legally* entered Chinese and Japanese markets since 1989, from the two CITES-sanctioned 'one-off' sales.

The first of these, agreed upon by CITES member countries in 1997, allowed Botswana, Namibia, and Zimbabwe to sell 50 tonnes of ivory – held in government stockpiles, and obtained from natural deaths and culling practices – to Japan. The sale took place in 1999, and it raised $5 million in revenue for the three countries. Then in 2007, CITES parties again agreed to allow the same three countries plus South Africa to sell 107 tonnes to Japan and China. This second sale, which took place in 2008, generated $15 million for the four southern African nations.

The idea was that these sales would flood Chinese and Japanese domestic markets with ivory, keeping the price low and therefore mitigating against future incentives to poach. Only it didn't quite work out like that.

Instead of a flood – releasing all the legally acquired ivory at once – China sold it to carvers in a trickle, and at greatly inflated prices. Along with the message that it is okay to buy ivory, this tactic immediately increased desire for ivory, by making it rare and expensive, at a time when more Chinese people had greater expendable incomes for buying luxury items. The conditions were ripe for an illegal black market to take off, and boy did it take off, as we have seen.

This is where the economics of the ivory trade becomes especially muddy. Some argue that it was the 1989 ban itself that drove up the price and status of ivory, by making the product rare and therefore more desirable. Perhaps this is the case, but recent analysis does show that ivory prices globally and within Asia were stable or falling between 1989 and the one-off sales.

More significantly, the rates of killing and poaching of elephants dropped off dramatically after the 1989 ban, and only increased substantially in 2008, *after* the two one-off sales.

Whilst some people still deny the link, or argue that we cannot draw conclusions on the basis of one coincidence in time, many anti-trade proponents argue that it was very simply the one-off sales that stoked demand, resulting in the terrifying rates of elephant poaching we have been seeing in the past decade or so.

At the same time as the 1989 ban was imposed, Kenya burnt its 12-tonne stockpile of seized ivory. I remember watching it on the news in awe as a child. This visually arresting event was meant to signal to the world that ivory was off-limits, that it was not a valuable, tradeable commodity any longer. Ivory does not burn easily, but Kenya wanted an image that would be difficult to forget. They hired Hollywood special effects pyrotechnicians to design a suitable mechanism, and the images are indeed iconic. And disturbing.

At the time, it was a controversial move – and remains so to date – with opponents arguing that destroying stockpiles only makes ivory rarer and so pushes prices up even higher, making poaching yet more lucrative and tempting. And whilst many other people agreed with Kenya's viewpoint, they found the burn itself very distressing, to think that these elephants had all been poached for nothing, with their ivory becoming a pile of ash.

Ignoring the controversy, Kenya remains resolute in its view that changing the perspective is key, that the world needs to stop viewing ivory as a commodity, and instead see it as something that belongs to elephants only. An increasing number of countries have since staged similar burns, or crushes of ivory. To date, nearly 250,000 kg of ivory have been crushed, burned, or both, in more than twenty different countries across Africa, Asia, North America, and Europe, with the single largest event happening in 2016, when Kenya set fire to its entire 105-tonne stockpile of seized ivory, representing an estimated 8,000 dead elephants.

Despite the coincidence in the one-off sale, rising ivory prices, and increased poaching – and the recent fall in ivory prices and reduction in poaching as demand has fallen – there are still many economists, politicians, and conservationists who are calling for the legalisation of international ivory trading, or at the very least for further 'one-off' sales.

Their argument goes that southern African states have robust elephant populations, as they have managed to protect their elephants from the poaching menace that has hit central and east African countries so badly. Because they are more 'successful' at protecting their elephants, they should be allowed to benefit economically from them – by selling further ivory stockpiles; using the money generated to fund more conservation initiatives and development projects to alleviate rural poverty.

How successful these countries are at conserving elephants is open to debate, with recent analyses from the Great Elephant Consensus team suggesting that whilst still relatively low, poaching rates are increasing in Botswana. Literally on the first day of my walk across that country, I came across a dead elephant carcass with its face hacked off on the banks of the Chobe river. The same rise has been reported in South Africa. But at its core, the debate on legalising trade in ivory is about generating the financial resources that are needed for conservation and rural development.

One analysis found that, when considered in the context of national economies, protected areas in Africa are more than a hundred times more expensive to manage than similar areas in

Europe, so the need to generate substantial funding for conservation in Africa could be a compelling argument.

The southern African nations that want to trade in ivory legally (still Botswana, Namibia, South Africa, Zimbabwe, and Zambia) argue that it is imperialist, condescending and racist for the rest of the world to dictate how they manage their 'healthy' elephant populations – especially when Europeans hunted our own large mammals to extinction in the past millennium. This kind of pejorative name-calling has gained more traction than you might imagine, as the underlying argument seems reasonable: shouldn't countries be allowed to make sovereign decisions about their own natural resources?

But the major problem with that line of argument is that elephants are not a static resource. They are living, breathing, moving animals that inhabit thirty-two other African nations, beyond the five that want to trade in them. Other nations that, to differing extents, do not have such stable populations, and some of which do not have the capacity to police further poaching epidemics.

There is no evidence to think that legal trade from southern Africa would result in the cessation of illegal ivory markets. In fact, I struggle to think of any legally traded products that do not have a parallel illegal trade. Arms, oil, diamonds, cigarettes, alcohol, prescription drugs, wood, pets, money, even music and films (with pirate DVDs now being replaced by illegal downloads and VPNs), all have thriving black markets or suffer from illegal products being laundered into legal markets.

Capacity and corruption are significant factors that would further limit the viability of any legal international trade in ivory. The temptation to exploit the system and launder illegal ivory into the market could be great, given that it would be all too easy to do. Fraudulent activity is conceivable at any and every link in the chain, from wildlife officials, transport-service providers, customs officials, retail-outlet staff, and even political players, each of whom can ease the path in return for some easy cash.

It would take only one or two 'bad apples' to allow massive quantities of illegal ivory into the market, and once it was there, it would be very difficult and costly to detect and remove, and the illegal trade could continue to flourish. The risk that legal trade could not be adequately regulated, policed, and controlled is great. Too great.

Yet there is one other, somewhat surprising source of ivory that is currently entering the market legal: that of mammoths! As the arctic tundra in northern Russia becomes more accessible – with permafrost melting and receding ever further – frozen, preserved mammoth carcasses are being found increasingly readily. The tusks can be legally removed and sold internationally because mammoths, as an extinct species, do not fall under CITES remit. But again, the problem is that it is extremely difficult to distinguish elephant ivory from mammoth ivory by eye alone, so already a lot of illegal elephant ivory is being relabelled as 'mammoth ivory' and sold without restriction.

Any increases in mammoth ivory trade, which could see considerable growth as people realise it is a legal way of buying and owning ivory, is likely to result in further laundering of illegal elephant tusks, and so push up poaching rates again. For this reason, Israel drafted a proposal to introduce mammoths as a CITES listed species – a controversial move for many, who argued how can an already *extinct* species be considered an 'endangered species'? The proposal was withdrawn from the last CITES meeting before it was debated, presumably because it was clear there was not much support for it, but this is an important issue that needs further consideration if we want to avoid any further increases in elephant poaching.

Legal trade simply does not extinguish illegal trade, and there is no economic, sociological, or biological reason to think that an ivory trade is going to be the exception. To what extent legalisation will increase demand for illegal ivory is perhaps more open to debate, but it will certainly not eradicate the demand. That means that any future legal trade from southern Africa is likely to further imperil – if not finally eliminate – elephants right across Africa.

It is for this reason that the largest coalition of countries lobbying and voting against the perpetual CITES proposals to permit sales from southern Africa are, in fact, other African nations. Kenya, Benin, and Burkino Faso are particular leaders, but thirty-two east, central, and west African countries – the 'African Elephant Coalition' – vote together on these issues at CITES meetings, vociferously opposing any further legal sales. Which also rather negates the imperialism arguments.

In a countermove, Botswana, Namibia and Zimbabwe, and possibly other southern African nations, are threatening to withdraw from CITES altogether, so they can resume trading in ivory (and other animal products) without restriction. Selling internationally would require their potential trading partners to also withdraw from CITES, which may or may not happen. But it is clear this issue is reaching boiling point.

How damaging would parallel legal and illegal trades be for elephants, if legal ivory trade was allowed, or the southern African nations did withdraw from CITES and resume trade? A recent analysis attempted to answer exactly this question by – unusually – looking not only at economic factors that influence trade, such as demand and sales, but also taking elephant demographics and biology into account. For example, the fact that there are two sexes of elephants, that these two sexes occur in different proportions and grow tusks at differing rates, and that the elephants with the most desirable tusks are the oldest and least numerous in any population. The results are not good.

The mathematical modelling showed that even with very low harvest rates, which would only satisfy low demand – and no parallel illegal trade or corruption – offtake soon became unsustainable and resulted in the extinction of the modelled population. For the population of 1,360 elephants used in the model, under the very best scenarios only 100–150 kg could be taken annually, and only if no poaching was occurring at the same time.

Given estimates of global consumer demand – based on poaching rates – and the current size of the continental elephant population, the perhaps achievable sustainable harvest rate of 100–150 kg of ivory from this population is at least three-to-six times lower than it would need to be from populations of this size to meet current demand.

In short, they showed that demand cannot be met with a sustainable ivory yield: any legal trade at current rates of demand will result in extinction.

One other form of trade that has been pursued by some southern African countries has been trade in *live* elephants – selling them to zoos in Asia, Europe, and the US. Zimbabwe has been particularly active in this regard, selling hundreds of calves to zoos in China. Namibia has made similar sales of fewer animals, as has Eswatini (formerly Swaziland). Many of these sales involve the capture of juveniles exclusively, which entails removing them from their families and sending them into captivity where they grow up without elders. The 'lucky' ones may get to live in the presence of an unknown, unrelated, older elephant, who has already been in captivity for many years.

The issue of keeping elephants in captivity could be a book in itself, suffice to say the practice comes with some fairly obvious welfare and ethical concerns. But the southern African states selling the elephants into captivity use a familiar defence: they have lots of elephants, and they need to raise money for expensive conservation programmes, so selling a few to fund the many is a necessary sacrifice. But not a sacrifice that much of the world supports any more, it would seem. At the CITES meeting in August 2019, the member countries voted in favour of essentially banning all live capture and removal of wild elephants into captive facilities.

Wild African elephants can now only be moved to other range-state countries, i.e. not outside Africa. Zoos in America, Asia, and Europe can no longer take elephants from the wild. You can probably imagine how Botswana, Eswatini, Namibia, and Zimbabwe feel about this loss of potential revenue. And how ecstatic welfare organisations are about the idea that no more elephants will be put through the trauma of being forcibly removed from their family and environment and shoved into a very confined space for the rest of their lives.

The illegal trade in ivory – and therefore elephant poaching – is not as easy to curb, but legalising trade is unlikely to be the silver bullet. The 183 countries that are party to CITES meet every three years to discuss and make amendments to trade deals in the 5,800 endangered animal and 30,000 endangered plant species that fall under its remit. Yet discussions at every meeting for the past two decades have been dominated by one or two species, of which elephants are arguably the most emotive and controversial. The same arguments and proposals are raised repeatedly, with pro- and anti-ivory trade sides refusing to budge.

Calls have been made for some kind of compromise, but it is hard to see what that could be, as any degree of legal ivory sale could have devastating consequences for elephants. More promisingly, I think, a change of focus is being suggested. Continued efforts to reduce ivory demand are necessary, to ensure the gradual decline in sales and poaching continue until they reach minimal levels that do not adversely affect elephant population numbers. But alternative, ethical ways of raising revenue to fund costly conservation programmes must also be developed, so that all African range states can better conserve the elephants that remain, and especially so that southern African states stop feeling hard done by.

Importantly, the livelihoods of people living alongside elephants in range states must be improved, and corruption tackled, so that the temptation to poach is greatly reduced. A longer-term perspective is necessary, which leaves polarising discussions about trading elephants and their ivory behind. As more people become aware of the consequences of the trade in illegal ivory, one would hope that elephants stand a chance of survival. But there's another threat to elephants that in this day and age is perhaps even more emotive to those leading the conversation: trophy hunting.

9

Taking Trophies

'Voortrekker'. It's an Afrikaans word meaning pioneer or path-finder – literally, 'fore-trekker'. It's the name given to the Dutch-descended settlers who trekked from the Eastern Cape into the interior regions of modern-day South Africa in the 1830s. It is also the name given to a male elephant living in Namibia's Kunene region, the coastal desert that encompasses the famously bleak Skeleton Coast. Voortrekker was one of only a handful of mature males left in this arid region, out of a total population of a few hundred elephants throughout Kunene.

Voortrekker was so named because he was one of the pioneer males who moved south back to the Ugab–Huab rivers area in the late 1990s, after the end of Namibia's War of Independence fought against South Africa. He had short, thick tusks that had cracked over his fifty-plus years, and a wizened, wrinkled head with the classic hour-glass shape of mature elephant bulls. According to local researchers, he was a very gentle and calm elder statesman. But Voortrekker was shot and killed by trophy hunters in June 2019, after being branded a 'problem animal' by authorities.

Voortrekker was accused of destroying water tanks and pipes around Omatjete village. The Namibian Ministry of Environment and Tourism stated that a hunter paid around $8,500 to shoot the animal, although it is not clear if this was the total hunting fee paid or merely the portion that was passed to the Omatjete community.

Members of the communities around Ugab apparently wrote to the MET after the hunting licence was issued, but before the hunt took place, saying that a mistake had been made: Voortrekker was an iconic elephant and photographic tourist attraction, and in fact he was *not* one of the elephants who had been destroying infrastructure in Omatjete, which lies to the east of Voortrekker's typical range. The Ugab community members wanted to find a solution to the problems that did not involve hunting and killing because, in their words, killing makes the problems worse, and they prefer to have calm and relaxed elephants in the area, not scared or aggressive ones.

Evidently, their appeals were not given much attention. Moreover, no official mechanism is in place in Namibia to ensure that individual 'problem' animals are correctly and accurately identified. Voortrekker – like many before him – was accused, labelled, and condemned on the basis of hearsay, with little actual evidence that he had really caused any damage. Part of the tragic irony is that Voortrekker had broken his tusks some weeks earlier, and so was a 'trophy' by reputation only, offering very little ivory.

Namibia's desert-adapted elephants are tall, thin, and with wider feet than other elephants – perfect for all that walking on sand in high temperatures. Analysis has shown they are not genetically distinct from other elephants, but they do show different behaviours that allow them to survive in the harsh desert region, such as digging in dry riverbeds to access underground water.

This specific knowledge about how to survive in the desert is likely learned and passed down through generations – youngsters learn survival tactics from watching their elders. These desert adaptations are pretty good candidates, then, for being the kind of social traditions – or cultural knowledge – that we mentioned in earlier chapters.

Through his life and death, Voortrekker embodies the debates and emotions that surround trophy hunting. Competition over space and resources creates conflict between people and 'problem' elephants, an issue we will explore more in the next chapter. His death also highlights questions about the ethics and sustainability of trophy hunting. But critical to understanding Voortrekker's fate are questions of value, ownership, and who should benefit from – and who should pay for – the remaining elephants that inhabit areas in which they evolved and have roamed for millions of years; an order of magnitude longer than the mere hundreds of thousands of years since the earliest *Homo sapiens* began inhabiting the same areas.

Killing elephants for trophies is big business. Some of it is legal, some of it not. Whilst poaching is illegal by definition, trophy hunts are usually legal, and controlled by strict guidelines on which animals can be killed and where. It may seem strange to conflate the effects of poaching and trophy hunting in one chapter, given that poaching causes death on an industrial scale, whereas trophy hunting kills only around a thousand elephants per year.

However, trophy hunters and poachers are aiming for the same thing – large tusks – and so target the same kinds of individuals: older, larger elephants who have more ivory. The people doing the killing, their motivations for doing so, and the reactions they garner, may all be very different, but the effects for the elephants of poaching and trophy hunting may not be so discernible.

Before we think about the effects that this killing has on the elephants, perhaps we should first deal with some of the contro-versies surrounding trophy hunting. Compared to the tens of thousands of elephants dying each year at the hands of poachers, the number of elephant deaths due to trophy hunting is small fry.

In 2017, the seven countries that permitted trophy hunting of elephants applied for 1,025 export licences between them,* and it is doubtful that all of these hunts would even have taken place. Countries often apply proactively for export permits each year from CITES, then those licences are eventually bought by the hunting tourists, so the number of export licences represents the maximum number of elephants that *could* be killed for trophies, rather than the actual number shot.

As of 2019, at the time of writing, Botswana also now permits trophy hunting of elephants, and they are auctioning eighty-six licences to Botswanan citizens, and seventy-two international licences, for hunts that will take place in 2020 (which is some way short of the 400 licences they used to apply for, before the short-lived 2014 ban on trophy hunting).

Yet despite the fraction of elephants being killed by trophy hunters compared to poachers, hunting garners huge media

---

* Those countries are Cameroon (which applied for licences to export the trophies from 80 elephants in 2017), Mozambique (25 licences), Namibia (90 licences), South Africa (150 licences), Tanzania (100 licences), Zambia (80 licences), and Zimbabwe (500 licences). Several of these countries have suffered significant declines in their elephant populations due to poaching in the last decade, most notably Cameroon, Mozambique, and Tanzania.

interest and it can stir up very strong emotions in even the most mild-mannered of observers. And – like poaching – it can still have profound impacts on the remaining elephants.

Those against hunting often use words such as 'barbaric' and 'inhumane', whilst those in favour might pitch it more as a noble sport; some hunters apparently enjoy baiting the opposition by posting grisly pictures on social media that show them standing proudly over their kills. Even many conservationists, who themselves have never hunted, can engage in vicious debates about the environmental benefits – or not – of trophy hunting.

When I walked across northern Botswana in the summer of 2019, I was following the annual migration of elephants in a westerly direction towards the Okavango Delta. I was travelling on foot through various national parks, so as well as Kane – my local San Bushman guide – it was necessary to have an armed wildlife warden along, too.

For the first few weeks I was joined by a man called Gareth Flemix, who was originally from South Africa but now lived in Botswana with his young family. Alongside photographic safaris, he had been granted honorary wardenship and was licensed to guide individuals on foot.

He was an ardent naturalist, who knew the sound of every creature in the bush by heart. He could name every single plant, tree and shrub that we encountered, and could follow the tracks of any animal across even the roughest terrain. He had lived almost his entire life in the bush and was passionate about wildlife and conservation, and he was never happier than when spying the trail of a buffalo or giraffe and sharing his love of nature with the tourists he looked after . . . he also shot a lot of elephants.

Gareth was a professional hunter, employed not only by the Botswana government to cull 'unmanageable' animals, but also by private trophy hunters. By his own admission, he had shot dozens of elephants, almost all of them bulls. I'd always found the concept of hunting big game somewhat distasteful, but Gareth seemed a reasonable, measured man and I was happy to listen to his side of the story.

One night, after a long day of walking through some of the most pristine wilderness Africa has to offer, we made camp under a baobab tree. In the distance I could hear the deep grumble of a lion and was glad to be sitting next to the campfire. I wanted to ask him what he thought of those people who were adamantly anti-hunting, and how he was able to justify what he did.

'Levison,' he said, clutching his rifle, which glinted against the flickering flames of the fire. 'You want to know what is the biggest threat to conservation?'

I said that I did.

He growled, 'We live in a world today where everyone has social media; you have people with opinions about what is going on in Africa, and yet they're not living here. On social media, suddenly everyone is a conservationist. If someone wants to have an opinion, they should base it on fact rather than emotions. We need to leave conservation to the real true conservationists; scientists, naturalists and wildlife biologists. Nobody goes to a doctor and tells the doctor what medicine he should prescribe. So why should people tell conservationists what they should do? They've studied their whole lives to protect this environment, so let's leave it up to them.'

With each breath he became more and more animated.

'Ten years ago, we had oceans of wildlife and islands of people, and today we have islands of wildlife and oceans of people, so of

course there's going to be conflict. Ultimately, it's vitally important we protect what we have left of this wonderful landscape.
And if that means we have to utilise a species to save it, then let's
do that.'

I interrupted. 'Okay,' I said, 'but tell me how you think that
killing animals saves them.'

'The truth is that hunting, if it gets done ethically and
correctly, is a tool of conservation. These large landscapes of
wilderness areas need money. They're not going to be left on
their own to be managed, because poaching will happen and
there will be more encroaching of cattle. Conservation needs
funds. Where do we get the money from? Not just from photographic safaris.'

He looked at me intently before carrying on.

'Only a very small marginal percentage of animals get hunted.
And that small marginal percentage of animals that you're taking
off per year brings in an extremely large, significant amount of
money. Sixty per cent of that goes back into conservation ...'

I was about to interrupt again, to ask how he could be sure
that was the case, but he seemed to read my mind and carried
on regardless.

'... but it needs to be managed. You need to ensure that
money is actually going back into conservation. The worst thing
about living on this planet right now is humans. We're untrust-
worthy, corrupt, we do things in different manners, and we
always have our own agenda. Hunting benefits conservation, by
bringing in a lot more funds that are necessary to protect these
wilderness areas. And I'm not pro-hunting if it isn't done
correctly and ethically. If it isn't done correctly and ethically, it
shouldn't even exist. And if it's corrupt, it must get shut down.
But conservation is a very complex scenario.

'You know, photographic safaris also have negative impacts. People drive through these villages with thousands of dollars of camera equipment; they don't even stop and wave at people, they go and camp on their land, and take pictures of those people's lions and elephants, and what are they contributing to their lives? Nothing, they're just driving through the village.

'Now put yourself in that man's shoes. His children are hungry, he can't plant maize because the elephants are eating his maize. He sees wealthy people driving through every day. What would you do? If your children were hungry? You'd go and shoot a buffalo, or impala, to feed your children. So, if those animals don't have an economic value to that man seeing the cars drive past, he's going to make his own decision to feed his family.

'It's a controversial issue, but it is crucial that communities benefit directly, not just indirectly from hunting and wildlife management. And if you have those two incorporated correctly then the villagers will understand.

'You know, a lot of people say hunting is bad. When hunting closed in Botswana, two years later the lion population decreased. Scientists came up with a figure, everyone was in shock, they couldn't believe it, they said it was impossible. It's simple: hunting stopped, those hunting concessions were empty, so cattle farmers went in. What does a cattle farmer do when he sees lions?'

Gareth paused, and watched the flames flicker.

'Lions and cattle don't sleep next to each other,' he began, then turned his head to look me in the eyes. 'Lions get shot. Lion populations decrease. Hunting accounts for a minute percentage of deaths of animals in Africa. Official culling, which is necessary for species management, is greater. So, if it's not hunting that's killing the animals, and it's not culling, what is destroying African wildlife?'

Before Gareth could continue, I suspected I already knew the answer.

'Habitat loss?'

'Correct,' he said, 'the loss of habitat. We need to focus on preserving habitat, not individual species, habitat. These animals need extremely large areas to roam. They have extremely diverse diets. They need large quantities of food source, and if we don't focus on protecting habitats, we'll lose it all. That is the crucial part of everything. People shouldn't get confused between the issues of hunting and culling. Leave that to the conservationists to make the correct decisions. Let the world focus on trying to protect habitat.'

But more on that in the next chapter.

Trophy hunting is a multi-million-dollar industry. A typical elephant hunt in South Africa may cost well over $50,000 – sometimes much more, depending on the duration of the hunt. Walter Palmer, the American dentist and keen hunter who controversially shot Cecil the lion in Zimbabwe in 2015, allegedly spent $54,000 on the trip, more than the average full-time annual salary in Britain (and considerably more than the average salary in Zimbabwe).

One study calculated that the total spent by trophy hunters in South Africa (where trophy hunting is permitted on privately owned land) amounts to $250 million a year. The Safari Club International, the biggest hunting organisation in the world with 50,000 members, proudly states on its website that it spent

$140 million on lobbying alone in the US since 2000 (that is, 'policy advocacy, litigation, and education for federal and state legislators').

For comparison, the entire budget expenditure of SANParks – the South African National Parks service, which manages all national parks within the country, employs thousands of people, and implements the country's conservation strategy – was about $120 million in 2019. People care deeply about hunting: about doing it, being allowed to do it, or about trying to stop people from doing it, and all sides dig deep into their pockets to further their cause.

Like the ivory trade, trophy hunting in Africa is a relic of the colonial era. It's almost impossible to think about trophy hunting without imagining Ernest Hemingway or Teddy Roosevelt-types in khaki slacks and shirt, with a rifle slung over the shoulder and pith helmet under the arm. And many of today's most famous safari destinations were set up to preserve hunting stock for the colonisers, or to protect the last few animals remaining after colonisers had all but wiped them out.

In the modern world, trophy hunters are the last vestiges of that legacy; typically, well-off Americans or Europeans (especially from Germany and Spain). Whereas elephant poachers tend to be local people, often poor and sometimes desperate. Discussing trophy hunting can quickly become uncomfortable, then, with the undercurrents of colonialism and racism that it can entail. But it's important to address the underlying issues.

Today trophy hunting occurs in a patchwork across Africa, being fully outlawed in some countries (such as Kenya, since 1977), permitted in certain areas and reserves (but not usually national parks) in other countries, or permitted for some species in some reserves in some countries (such as Zambia, which had

An orphan at the David Sheldrick elephant orphanage in Nairobi, Kenya. These calves are taught to feed themselves whilst being rehabilitated.

Baby elephant. Calves stay close to their mothers for almost a decade, learning new skills daily.

Tim, one of Africa's last remaining 'big tuskers' in Amboseli, Kenya.

Fang Ivory traders in West Africa, late 19th century.

Ernest Hemingway on an elephant hunt.

A poached elephant, Kenya. The ivory trade is a gruesome business.

Kane Motswana, my San Bushman guide, holds a 45kg tusk, from a recent find in Botswana.

Ivory ornaments and trinkets for sale in Beijing, China.

A modern hunt. I join Mike Chase and Larry Patterson of 'Elephants Without Borders' to collar a bull so that it may be tracked and protected.

A member of the Botswana Defence Force with seized ivory.
Poaching kills 20,000 African elephants a year.

Kenya burning its stockpile of ivory in a show of defiance
against those wishing to keep the trade alive.

Efforts to reduce human–elephant conflict include making 'chilli bombs' which scare elephants away from farmers' fields.

Deforestation in Uganda. Habitat loss is the biggest threat towards the survival of elephants.

More elephants come into contact with humans as the human population increases and elephant migration routes are disrupted.

Elephants' crossing. human–elephant conflict is on the rise as elephant habitats face more encroachment by humans.

Author and Kane Motswana with ancient San rock art showing elephants in Botswana.

The author with Molelo the orphan in Botswana.

a ban in place until 2015 on big-cat hunting, but hunting other species was permitted). Other countries, such as Botswana and Tanzania, oscillate between outright bans on hunting, and reinstating 'sustainable use' practices, depending on the political and environmental winds of the time.

Sustainable use is the phrase of the moment. Think of cutting down trees of a certain size and species for firewood or building or other infrastructure, whilst ensuring that trees not fitting the species and size regulations are left alone, and that replacement trees are planted and saplings cared for. That is sustainable use. Cutting down any and all trees in any area you feel like, with no actions to ensure timely replacement, is not sustainable. It is simply destructive use with only short-term gains.

Whilst the term sustainable use can be applied to any natural resource, it is very often used as a euphemism for trophy hunting, as the term inherently implies the conservation benefits that trophy hunting can have.

Proponents of trophy hunting argue it offers three major benefits for wildlife. It generates incentives that encourage wildlife conservation; it generates financial revenue to support this conservation; and it preserves habitat for wildlife, thereby reducing conflict and negative activities such as poaching. These benefits combine, according to advocates, to allow increases in population numbers of endangered species such as elephants, some of whom will also become trophies one day.

It must be said that it is true that southern African countries who permit trophy hunting do have some of the most robust elephant populations (and lowest rates of poaching) on the continent. But we should not forget that the snapshot of population numbers at any one time tells only part of the story. Elephant numbers are rising in South Africa because they were so critically low a century ago; numbers are high in Botswana because of an influx of 'refugee elephants'; and whilst numbers may be high in Hwange, big losses have occurred in other areas of Zimbabwe, due to poaching. It's a difficult numbers game to play.

Analysis has suggested trophy hunting adds some $425 million to African economies, and supports tens of thousands of jobs, which – if true – would be hard to object to. But this report was commissioned by the Safari Club International, and so may not be entirely objective. Other analyses (commissioned by equally partisan bodies, such as the International Fund for Animal Welfare) suggest the figure is more like $132 million (so still pretty large, but much lower in terms of contribution to GDP of the countries that permit hunting.) The real number is probably somewhere in the middle.

Anyway, most revenue remains within the private sector, and so is not necessarily ploughed back into conservation efforts, and the number of jobs created and maintained by the hunting industry may be much lower – somewhere in the range of 7,500 to 15,000 across the continent, not the 50,000 estimated by SCI. Additionally, the percentage of hunting revenue that actually reaches communities living next to elephants and other wildlife may be very low, even where community-based ownership projects are in place, and so may not always be sufficient to maintain the 'goodwill' and incentives for wildlife protection.

When it is well managed, trophy hunting probably does contribute to conservation objectives. This is particularly true in poor areas where local people live alongside – and often in conflict with – wildlife, and who would otherwise gain no material benefit from having to share their homes, land, fields, and water sources with elephants.

A lot of research shows that employment and income generated from trophy hunting – when shared equitably – promotes conservation values in the people living alongside the animals. Ideally, the people see a reason to look after wildlife as the hunting of animals gives them income, so they can afford to be more tolerant if the animals eat crops, destroy fences, or consume all the water and grass that would otherwise have fed their livestock.

But unfortunately, there are all too many examples of trophy hunting not being well managed. Sometimes local people do not receive any or all of the money they have been promised; or else the hunting quotas are set entirely for economic gain, without considering conservation objectives and what rates of offtake would really be sustainable; on occasion even bad hunting practices occur – where animals are wounded but not killed, or witness the violent deaths of their group mates – which can make the animals even more aggressive or dangerous to local people. This last issue may be especially problematic when it comes to elephants: I have spoken to quite a few Maasai in Kenya who are totally convinced that elephants kill out of retribution.

Elephants will normally try to avoid Maasai and their livestock – moving away from an area if they hear the familiar cow bells, shouts of conversation, or catch the distinctive scent of the herders. This is because elephants are occasionally speared by Maasai warriors, for example if there is conflict over access to water in the dry season. However, many warriors I know have stories of groups of elephants returning and literally seeking out and hunting down a cow or goat, then goring it to death, after one of their own has been speared.

This kind of escalation of conflict does neither side any favours, as we shall see in the next chapter, but it may illustrate yet another remarkable thing about elephant minds – a desire for revenge!

Of course, hunting should not only be about economic benefits, which may or may not be exaggerated. We should also think about the well-being of the communities living alongside elephants. One case study has been published about the Nyae Nyae conservancy in the Kalahari desert of Namibia, which allows trophy hunting under the 'Community Based Natural Resources Management' model, whereby hunts are operated by an external body, who submit tenders for the licences to run commercial hunts on community-owned land. The study suggests that local San people have been bullied, bribed, overworked and underpaid, and generally disrespected by some operators.

Reports such as this somewhat negate the now familiar argument about sovereignty that has been lodged for trophy hunting, too: nations should be allowed to decide their own conservation and management strategies for the animals living within their borders, and those of us from countries far away, who do not have to live with or pay for conserving such species, should mind our own business. Judgements about what is ethical and 'right' may seem very different from an armchair in Britain, compared to a smallholder's field of trampled and destroyed crops in Zambia. Yet it seems that complaints about the lack of thought for, and patronising or racist attitudes towards, local communities may well be levied against some members of the pro-hunting lobby as well as anti-hunters.

While it is clear that we must take the viewpoints of people living alongside elephants into account, it is extremely difficult to make informed and balanced opinions about the social and economic benefits of trophy hunting – as it is with poaching. But a real-life example illustrates some of the problems.

In some areas, such as the privately owned nature reserves that share an open border with Kruger National Park in South Africa, mixed tourism models coincide, with both photographic and hunting safari operators sharing the reserves. Of course, the operators and managers of these reserves try very hard to keep the two kinds of tourist separate. I suspect many of the photographic-only tourists to this area are entirely unaware that the animals they are photographing could be shot by high-paying trophy hunters; and even if they do know, they probably don't want to witness any animals being shot by the hunters. Yet this is exactly what happened in a reserve in late 2018.

A group of trophy hunters targeted a male elephant, aged between twenty and thirty, and had followed him to within about 800 metres of an eco-tourism lodge – within sight of a group of four tourists on the viewing deck of the lodge. Despite unwritten agreements about not shooting when close to lodges, the hunting party took aim anyway. Bad aim, apparently, as it took several shots (thirteen, according to the eyewitnesses) to kill the elephant, who ran away trumpeting and screaming after the first shot rang out.

No formal charges have been brought against the hunters, as nothing illegal occurred. But I do wonder if this incident – well publicised in South Africa at the time – has had a negative impact on the number of photographic safari-goers choosing to visit this reserve. If so, the jobs of many people within the huge photo-tourism industry of the area could end up being vulnerable.

I remember going to the Kruger National Park in South Africa and speaking to local villagers there. It was rather pessimistic. Most of the people I met said that tourism and the high-end lodges that brought in hunters and photographers has done little to uplift them. In Botswana, it was the same. Many felt excluded from and undervalued by the nature reserves. And this is in the two countries with some of the best tourism infrastructure and economic planning in the continent. Clearly, balancing the needs of tourists, hunters, landowners, and community members is very difficult indeed.

Perhaps the most compelling argument in favour of trophy hunting is that it maintains wild habitat that would otherwise be developed or converted to agricultural or pastoral land, resulting in further loss of habitat for elephants. Over one million square kilometres of land across Africa is currently dedicated to trophy hunting, which is more than is protected by national parks. If this were to be lost, then that would inevitably be a huge blow.

The threat of habitat loss is something I'll come onto in the next chapter, but it's something to bear in mind. It is because of this threat of habitat loss that few conservationists, whether they support hunting or not, would advocate for an immediate and total ban in hunting. Alternative means of funding wild areas have to found.

In an ideal world, changes away from hunting might offer opportunities to find potentially more lucrative ways to fund conservation and maintain local and governmental incentives for keeping land wild. Of course, additional eco/photo tourism can pick up some of the slack – but it is predicted not all of it, especially as many hunting concessions are in very remote and wild areas that might not be attractive to 'luxury' safari tourists. As more ideas are generated, the debate about trophy hunting is intensifying, and perhaps opening up to new possibilities. It all depends on whether there is the will to create that 'ideal world'.

You may be convinced, or not, by the economic arguments in favour of trophy hunting. And you may agree, or not, that 'hunting is natural' – a frequent argument proposed in favour of the sport. But these arguments all seek to rationalise the killing of elephants entirely from a human perspective. What about the elephants, themselves? Is hunting biologically sound – does the *sustainable* in 'sustainable utilisation' really hold up? Or does the killing have any implications for elephant populations that need to be factored into our thinking, beyond the ethics and economics?

A key argument against trophy hunting of elephants is that increasingly it is failing to meet the industry's own sustainability criteria. Trophy hunting – like poaching – is inherently biased towards the larger, older males, as these are the individuals with the largest tusks. And so many large males have now been killed that average tusk size has decreased significantly, even in the past decade.

An analysis of trophy sizes taken from hunts in the Matetsi Safari Area of north-west Zimbabwe showed all tusk trophies taken were below SCI minimum scores, and declined year on year. The age of the males being shot remained constant, so declining trophy size was not due to taking younger males, but rather that the males who were left to shoot simply had smaller tusks: large-tusked or real 'trophy' males had disappeared. If hunting really is sustainable, the disappearance of trophy animals simply should not happen.

Given the location of this population in north-west Zimbabwe, poaching could have contributed to the loss of large-tusked males in addition to the trophy hunts. But consider the elephant population that lives in Mapungubwe Transfrontier Conservation Area, which straddles the borders of Botswana, South Africa and Zimbabwe, and where there is relatively little poaching. Each individual country may attempt to impose sustainable hunting criteria, but there is currently little coordination between the three countries on the hunting quotas they set – the upper limit of how many animals can be shot. This means the population is essentially under three times the hunting pressure it would be if it fell entirely within one country.

The population may be able to sustain losing say eight or nine elephants per year to trophy hunting, but with each country setting its own quota, many more than this are being killed, entirely legally. In fact, it was predicted in 2013 that if the current rates of hunting continued, trophy males will completely disappear from the population by 2022. Until cross-border, coordinated authorities are established that take elephant biology and ranging patterns into account, 'sustainable' quotas are wide open to failure.

Entirely losing large-tusked male elephants means that population structures are being eroded and important genetic diversity is being lost. We do not yet know enough about the genetics and heritability of tusk size, but selective killing of large-tusked males that would normally be responsible for much of the breeding is obviously going to alter the genetic make-up of a population, which could have knock-on effects that we haven't yet imagined.

On top of that, given what we know about the close social-grouping and life-long learning of both male and female elephants, it would be crazy to think that we can keep killing elephants with no impact on their behaviour and biology. A primary justification of trophy hunters is that they typically remove only the oldest individuals from a population – those with the biggest tusks, horns or mane or whatever the prize may be – and, crucially, that these old individuals are no longer active reproductively. So removing them does not have any serious effects on subsequent generations, which is a key part of what makes hunting sustainable. That argument may work for buffalo or antelope, and perhaps even for lions, but for elephants, it categorically does not stand up.

For elephants, hunting (and poaching) simply cannot be justified by stating that the animals taken are of no further use to the social group or population they came from. Remember that females prefer to mate with bigger, older, wiser males, who are proving their fitness simply by being around and in musth. Paternity data gathered from multiple generations of elephants in Amboseli confirms this, showing that males father a low but increasing number of calves from their first sneaky matings until their early forties, but they are in their reproductive prime from their forties until their mid-fifties. Older males do not stop fathering calves, and even the oldest bulls remain very important to elephant reproduction.

So, put bluntly, it is a complete fallacy that old, 'trophy' male elephants are no longer contributing to elephant society, and any claims to the contrary can and should be dismissed immediately. I read something recently that stated an old male elephant who had been shot by a trophy hunter in eastern Namibia was on his sixth and final set of molar teeth, and that these were 'half ground down already', as though that justified his fate. Bear in mind that the sixth set of molars erupt at around thirty years of age, and the life expectancy of an adult male can be up to sixty-five years. So that final set of teeth can last over thirty years.

To be half worn down implies (if we assume a regular rate of wear) that the elephant who was shot was around forty-five years old – at the beginning of his prime breeding period. He could have continued fathering calves for another fifteen or twenty years. That means he could have successfully sired twelve additional calves if he had not been shot. He was hardly at the end of his reproductive life!

It is not simply a case of sacrificing one elephant for the greater good, because older male elephants do not merely contribute sperm to society. Just as older females contribute knowledge and help to maintain normal functioning society, older males are equally important in this regard. In the 1980s, multiple 'translocation' operations were undertaken to move about 6,000 animals of various species into a newly proclaimed wildlife reserve called Pilanesberg, in the North West Province of South Africa, to restore the area to a natural wilderness. Among these 6,000 animals were forty-four elephants – twenty-seven males and seventeen females. All forty-four of these elephants were under ten years old – yes, they were all juveniles. There were no adult elephants in the population. Are alarm bells ringing? Then in 1993, an additional thirty-six juveniles were added to the Pilanesberg gang.

The elephants had been sourced mostly from Kruger National Park, where culling of adult elephants was still ongoing at the time; juveniles were 'spared', and sold off to bidders to either establish elephant populations in new reserves, or to be trained for captive tourism facilities that offer activities such as elephant riding. The juveniles that moved to Pilanesberg seemed to settle in okay, and things were pretty quiet for the first decade. By 1989, some females began to have calves of their own (so males must have been fathering calves at around age eighteen), and the population began to grow.

But in 1992, a decade after the elephants had arrived, park management found a dead rhino. Then another, and another. Not just dead, but mutilated. It didn't fit with poaching, as the horns were still intact on the rhino bodies, but the deaths were evidently gory and unpleasant. Between 1992 and 1997, more than forty rhinos were found dead like this. After much investigating and debate, it became clear that the lacerations on the carcasses were caused by tusks: elephants were responsible for the rhino killings!

It was confirmed with remote cameras and radio collars fitted to rhinos that a group of several now adolescent male elephants was carrying out the attacks. Even more surprisingly to biologists, these teenagers were showing signs of being in musth – that prolonged testosterone surge that male elephants show annually, usually only from their mid-twenties or thirties onwards.

These young males were coming into musth far too early, and apparently losing their minds in the process. A group of hormonal adolescents with no older guiding influences went totally berserk, killing 3-tonne animals in testosterone-fuelled rages. It sounds like science fiction, but it was tragically real. In fact, it also happened at several other reserves in South Africa that founded elephant populations with juveniles only.

Park managers shot a few of the known culprits, including two male elephants who killed a tourist in 1993, but this did not eradicate the core of the problem; other males simply came into musth as well, and turned into angry, hormonal rhino killers. Eventually it was decided that these aggressive, cocky adolescents had to be put back in their place, and that the most effective way of doing so would be to introduce some real men.

In spring 1998, six mature male elephants from Kruger were introduced to Pilanesberg. The largest was a male named Amarula, in his mid-forties. One report suggests the younger males immediately flocked to Amarula, but one adolescent in musth apparently didn't pay sufficient respect. Amarula hit him in the stomach so hard the youngster was knocked of his legs, flying several feet into the air. The writing was on the wall: a new boss was in town. Musth states in the youngsters immediately declined and the rhino killing stopped.

These elephants were actually disrupted by culling, not trophy hunting or poaching, but the effect is often the same − male elephants growing up in the absence of larger, older males can be socially dysfunctional. They become unable to deal appropriately with the hormonal surges they experience prematurely, causing great damage to their environment. Older males serve an important function as role models who maintain order and decent social behaviour. The loss of older elephants − the ones that are most targeted by trophy hunters and poachers alike − can have disastrous consequences for the elephants (and other animals) left behind.

Of course, none of this means that we should be shooting younger males instead. Younger males are obviously necessary for becoming the elders of tomorrow, but they also serve an important role today. Studies of elephant movement patterns, recorded from the kinds of tracking collar that Mike Chase fits, show that it is the younger males who are the explorers of elephant society. When elephants need to move out of old areas

– perhaps because they are being persecuted, or climate patterns are changing and they need new places to feed and drink, or maybe new areas open up to them after fences are removed – it is the younger males who go first and explore.

Changing rainfall patterns across Botswana mean that the Okavango Delta is becoming slowly drier, whilst the Boteti river that empties into the Makgadikgadi Pans area started flowing again in 2009 after being dry for eighteen years. This change in water movements has been followed by a change in elephant movements, with a few, then several, then hundreds of elephants moving into the Makgadikgadi area after a long absence.

As I discovered when I walked across the pans in 2019, the first elephants to arrive were the 'explorers' – young males, just as Voortrekker was among the first to return to Ugab, when he was in his mid-twenties, followed by older males. I didn't see any myself, but the tracks were there – an incredible zig-zagging line straight across the glistening white desert, linking the 'islands' where they presumably stopped off to feed on the vegetation.

The family groups followed later on. Without these pioneer young males to explore and encourage movement into new areas, elephant populations could get stuck in dangerous or inhospitable ground, which would likely hasten their death.

All animals adapt to their environments, or else they die. That is the basic essence underlying evolution by natural selection. In evolutionary terms, we tend to think of adaptations being genetic mutations, but animals can also learn to behave in particular ways that enable them to fit into and survive in a particular habitat. If the animals who have learnt a strategy pass that behaviour on to their offspring or others in their social group – because the others learnt from watching the experienced individuals do it – behavioural traditions or 'cultures' can be established.

Think of Voortrekker, the old male in the desert population of Namibia. As an experienced pioneer, Voortrekker would have held invaluable knowledge about water sources and food locations spread across a vast, arid area. It was knowledge such as this that allowed elephants to survive a severe drought in the region in 1981, whilst 80 per cent of other large herbivores such as zebra, kudus, and springboks, perished. And it is knowledge that the remaining elephants in the Ugab region may not have yet fully acquired from him. His death, and the associated loss of his decades of knowledge, could have knock-on effects for this fragile population that may only become apparent in drought years to come.

Namibia rightly argues that this desert population is not genetically distinct from other savannah elephants, but that does not mean their conclusion that these elephants are therefore no different to any others is correct. I am not genetically distinct from a San Bushman, but I am pretty sure I couldn't survive in the desert for too long on my own, like they can.

It is the acquired cultural knowledge of these desert elephants that sets them apart and allows them to survive in this difficult habitat. They may not be genetically distinct, but they are *culturally significant*, and allowing them to be hunted to the point of local extinction, which certainly seems possible at the moment, given the rate at which males are being removed from this population, risks losing not only elephant numbers, but centuries of acquired cultural knowledge.

The loss of knowledge and alteration of populations and genetic make-up that results from trophy hunting of elephants means that – like poaching – it is not sustainable in its current form. Trophy hunting and poaching both drastically alter the social and genetic make-up of elephant populations, and both have significant implications for the behaviour and ecology of the individuals that remain.

However, it is not solely about the animal – we also have to think of the habitat, other wildlife and the people who share their land and livelihoods with elephants. Discussions about trophy hunting quickly become discussions about economics and socio-political issues. Is it perhaps okay to kill an elephant if it means that it supports people financially – providing jobs and income and infrastructure that would otherwise be missing? Or if it contributes to the maintenance and protection of wild habitats and ecosystems?

Maybe the loss of individual elephants, knowledge, or traditions is justified by the greater good for society and environment. However, it's fair to ask why it is okay for that loss to be perpetrated by foreigners with loads of money and privilege who want to take tusks away as a personal trophy – but it is not okay to remove tusks for a person who has spent much of their life living alongside these animals; struggling when the elephants have eaten their crops, or blocked their roads so they cannot walk to school or work.

Hunters are lauded in some circles as contributors to conservation, whilst poachers are vilified and face enormous fines or heavy prison sentences in many African countries. Of course, the root of this difference lies in the ownership of natural resources. Poaching may benefit the poacher and his family or associates, but revenue channels are lost for the 'owners' of the elephants. Professional hunting companies, owners of hunting concession land, and the local and national government all lose potential income when an elephant is poached. One is stealing, while the other is capitalism.

I don't mean to imply I have *The Answer*. Or even *any* answers to these hugely complex social, political, economic, and environmental questions. Evidently trophy hunting is far more complex than simply deciding what you think about shooting animals. Even so, wherever you land in these debates, there are biological facts that are inescapable – but often ignored.

Hunting of some species may well be sustainable, but we should remember that elephants are intelligent living beings with complex social structures. We can no longer pretend that shooting them at will – be it under legal or illegal guises – can continue for as long as we want with no consequences. The selective removal of larger, older elephants for both legal trophy hunting and illegal poaching has intense and severe consequences for the remaining elephants, and for the wider habitat, ecosystems, and people sharing the areas.

However, the furore over trophy hunting in some ways sadly misses the point. Even the killing of elephants for ivory, while it may have been the underlying cause for most previous deaths of elephants, is no longer the biggest threat.

By far and away a bigger risk to the survival of elephants is a more subtle type of destruction. As the land in Africa becomes ever more fragmented by roads and changed into farmland, the savannah and forests in which elephants have roamed for millennia are being constantly eaten away. Every week that passes signals the not-so-gradual annihilation of the elephant's ancestral home, and this could signal the death knell of those ever-shrinking wilderness areas that elephants need in which to be safe.

This indirect effect of human interference is the greatest threat to the majority of wildlife across the globe, and it is us humans that are the root cause of that loss, due to our ever-increasing need for resources.

10

The Elephant in the Room: Habitat Loss

In the autumn of 2019, I travelled to the Democratic Republic of Congo on a photographic assignment as a guest of the team at Virunga National Park, the oldest reserve in the country. Virunga appears from above as a long strip of tropical rainforest along the border with Rwanda and Uganda. It's filled with towering volcanos and misty mountains and is famously home to not only a few hundred of the last remaining mountain gorillas, but also one of the most vulnerable and isolated populations of forest elephants left in Central Africa.

An hour and a half's drive north of the city of Goma is Kibumba camp, a collection of canvas tents in a clearing on the slopes of Mount Mikeno surrounded by dense forest. It was a spectacular sight and unlike anywhere I'd been before. In the distance, thunder rumbled as dark rainclouds melted seamlessly into the billowing smoke from the mighty Nyiragonga volcano, which rose from the jungle valleys like a giant pyramid.

I'd finally fulfilled a lifelong ambition to visit the Congo, inspired ever since my grandfather's stories of the elephant graveyard as a child. Under the watchful eye of half a dozen armed rangers, who protect tourists from sporadic banditry and kidnapping, I felt a sense of excitement as I found myself trekking deeper into the jungle in search of the magnificent great apes.

The rangers in Virunga have a special relationship to their charges. They go out every single day to track the gorillas and many of the apes are named after wildlife officers who have been killed by poachers and armed gangs. Many of the gorilla families in Virunga are habituated and used to humans, thanks to the rangers' dedication, which meant that it didn't take us long to find them, because the rangers knew where they liked to feed and sleep. Despite the cold chill and incessant rain, I whiled away one of the most exhilarating hours I have ever had, watching these beautiful creatures play, eat and frolic in their natural habitat. It was remarkable to be so close to these wonderful animals and I felt very lucky to be in the presence of such a rare thing. But I wasn't here just for the gorillas. I wanted to be able to track forest elephants, too.

I had dreamt of this moment for decades, to hack through the volcano forests of my childhood imagination in search of the mystical elephant graveyard. Even though I knew it was fantasy, simply to be here, in the ancient jungles where apes and elephants meet, was magical. Regardless of the chances of actually seeing a forest elephant, I was determined to try.

Benoit, one of the rangers who was guiding the way, burst out laughing at my suggestion. 'If you want elephants, go and see savannah elephants on the Garamba plains. These forest types are impossible to find.'

He wasn't wrong. We spent two full days of solid jungle hacking, covering over forty kilometres and having to chop new paths through the dense foliage – and the best we got was spotting a few piles of week-old dung and some footprints around a mud pit. No elephant graveyard; no tusks, not even a few bones, and certainly no live ones to be found.

None of the rangers knew how many elephants had survived in this isolated patch of jungle, but it couldn't be many, and as I returned to camp, admittedly a little disappointed, I went to stand on the edge of the hillock overlooking the valley.

Despite its wild appearances and the lush, fertile expanse of seemingly endless greenery, I could hear the noise of hundreds of voices; humans, chattering away in the distant villages. Smoke from countless fires spirited upwards from invisible homesteads, and to the east, in Rwanda, I noticed an almost vertical mountain slope had been hacked into terraced fields. Impressive though it was as a work of agriculture, I felt a deep sense of desperation for the future. What chance did the elephants have, if this relentless tide of human growth continues?

The world's human population has doubled in the last fifty years. In Africa, it has more than trebled, rising from 350 million in 1970 to 1.3 billion people now. Africa's population will double again in the next thirty years, and it will soar to 4.4 billion people by the end of the century. The global economy has grown four-fold since the 1960s, and global trade has seen a ten-fold increase. All of this means our demands for water, food, shelter, materials and energy have increased exponentially, and will continue to intensify. Rapidly.

The obvious consequence of having more people in the world, taking up more space and using more resources, is that there is less room for wildlife. We humans have now modified more than half of the earth's land surface in one way or another. So as human populations and needs increase, wildlife will become ever more marginalised.

People have changed the use and structure of this land, which in turn has altered the habitats and had a dramatic impact on the balance of wildlife and ecosystems – the net result being a loss of biodiversity. The biomass (the total weight of living organisms) of *all* wild mammals globally is less than a tenth of our own biomass. The total global weight of our livestock, however, is nearly double that of humans; we are filling the world with cows, sheep, goats, and chickens. And this all comes at the expense of wildlife.

It is predicted that thousands of wild mammal, bird, reptile, and amphibian species will lose their natural habitats in the next fifty years, and so become critically endangered. For the past century or so, protected areas have been the principle solution to counteract biodiversity losses, but the rapid rate of human population growth means there is often encroachment into these parks or reserves. It has been estimated that a third of all protected areas around the globe are under intense pressure from human use. 'Fortress conservation' – keeping people out of reserves – is crumbling.

It is no surprise, then, that elephant range across Africa has decreased from an estimated 7.3 million square kilometres in 1979 to just 3 million square kilometres today. Loss, change, or encroachment of elephant habitat can take many forms. Local-scale encroachment by subsistence-level farmers is one end of the problem, with more industrial-level intrusions at the other.

Biologists often talk of 'carrying capacity'. Simply put, it's the number of animals that one defined habitat area or ecological system can support without becoming degraded. A given area has physical space for a finite number of plant species, and these plants can feed only a certain number of animal species without being eaten into local extinction, and these herbivorous animals in turn can support a precise number of predators, and so it goes on right up the food chain.

Degrading habitat quality – reducing the amount of plant food species available, for example by allowing overgrazing by livestock, or planting single agricultural crops – decreases the carrying capacity of an area. So, in effect, when biologists talk about decreased carrying capacity, they are essentially talking about habitat loss: a reduction in the amount of land and space available for a species.

Continentally, habitat loss is a major threat to elephants, who need to feed over vast areas to meet their nutritional requirements. The 1979 population of 1.3 million elephants covered a range of about 7.3 million square kilometres. By 1989, the much smaller population of 650,000 elephants covered around 5.9 million square kilometres. And today's elephant population of 415,000 animals resides in about only 3 million square kilometres across the continent.

Elephant range has halved in thirty years. The *potential* range of elephants in Africa is around 20 million square kilometres, meaning elephants are currently found in only 15 per cent of their potential range. And of the areas where they are found, less than a third is designated as a park or reserve.

We have a clear problem here. Only 30 per cent of the elephants' current range is protected, but 84 per cent of elephants (based on Great Elephant Census figures) range *inside* protected areas. This means they are crowding into a very limited area, presumably where they feel safe.

Moreover, in the same areas as the census counted 352,000 elephants (lower than the total continental figure of 415,000, as remember the census only aimed to count 90 per cent of savannah elephants, and no forest elephants), more than three million head of livestock were also found. Cows, sheep and goats outnumber elephants by ten to one, in prime elephant habitat.

You have only to look at satellite images of the level of deforestation in central African countries such as Uganda and DRC to see how much indigenous rainforest has been lost because of sugar cane and palm oil. Or read any local African newspaper on any given day to see the scale of habitat loss: hydroelectric dams, exploratory oil drilling or mining extraction, heavy logging, and road building, all occurring within elephant habitat *inside* protected areas, very often *with* national government backing.

Even when we are not digging up elephant habitat, we can be changing and polluting it, as we are with the rest of the world; with plastic, pesticides, chemicals from industrial waste. And of course, with livestock.

Elephant density – the number of elephants that can be found in any one area – should be related to resources such as water and food availability. It's biology 101: animals can be found near their sources of food and water. But it has now been shown in several different locations across savannah elephant range that elephant density is best predicted by the presence – or rather the absence – of humans.

Elephants try to avoid humans as much as possible. There is a threshold level of human activity, above which elephants disappear from an area: when any human use reaches about 40–50 per cent of the activity within an area, elephants would prefer to move away to less disturbed habitats. Interestingly, at a continental scale, as the literacy rates and GDP of human populations increase, so does the number of elephants, which makes sense as we know that education levels affect attitudes to conservation. But the flip side: as corruption rates increase, elephant density decreases.

Elephants would prefer to avoid people, but this is not always possible. And when elephants have to live alongside humans – overlapping with them in space, and perhaps trying to use the same resources such as a source of water – conflicts can occur. According to the World Bank, many of the world's poorest people (those living on less than $1.25 per day) live in rural areas of African elephant range countries, so the people elephants meet and have conflict with are often exactly those people who are already struggling: subsistence-level farmers eking out a living in remote areas.

To many farmers, agriculturalists and pastoralists, elephants represent a distinct threat to livelihood and life. Elephants can cause death, injury, and financial loss to humans. People may see livestock killed and crops eaten or destroyed, or have their daily activities curbed by elephants blocking their path to school, to their employment, to their fields, or to places where they would collect necessary water or firewood.

But of course, these same farmers also represent a threat to wildlife, with elephants having to compete with livestock over access to food, being chased away from water sources or feeding grounds, and perhaps being speared, shot at or poisoned. When people lose their crops or livestock, it can often result in retaliatory killings of elephants – many elephants die in this way every year; even though it can be dangerous for farmers to defend their land against them, with many people also dying in such encounters. Rates of injury or death to humans and elephants are known to increase in times of drought, when competition for resources – the elephant's desire to access them, and the farmer's need to defend them – is at a peak.

None of this is meant to imply that the people who are encroaching into elephant and wildlife habitat are necessarily 'bad', it is purely that it's happening, and nobody seems to want to talk about it. Whether at the level of the local farmer looking for space to plant a few crops or raise goats, or governments or corporations needing to expand energy or resource production to feed and fund whole populations, they are simply people trying to make a living.

It is natural and necessary for the nations of Africa to develop and change land for the benefit of their people, and it would be hypocritical of rich Westerners to object. Consider an analysis by the World Wildlife Fund, which showed that if the entire global population was as developed and consumed as much as we do in the high-income 'global north' (which essentially comprises western Europe, Israel, Russia, Japan, Singapore, Australia, New Zealand, the US, and Canada), we would need three planets to sustain everyone! Given our rates of consumption, we cannot begrudge the development of African nations. But we can hope it is achieved in a more sustainable way than our own development.

The Republic of Botswana is held up as a beacon of such development within Africa. The continent's oldest continuous democracy, it has a good record of free and fair elections in a multi-party system since its independence from the UK in 1966. With a steadily growing economy and a relatively small population of around 2.3 million people, poverty has been managed pretty well and the country is able to provide a modest, universal old-age pension, which is rare in Africa. Levels of life expectancy, education, and per capita income – factors which together make up the 'Human Development Index' – are among the highest in the continent, with only the Seychelles, Mauritius, Algeria and Tunisia doing better.

Yet, like all countries, Botswana has its problems: inequality does remain, there's high unemployment, and shocking poverty still exists. Some minority San people have been forcibly removed from their historic land and relocated into defined reservations, where unemployment is especially high. Moreover, rainfall is often scant, making water a very precious resource. Drought and desertification are environmental problems that are on the rise, exacerbated by overgrazing by the several million cattle that help to support the economy, making agriculture and pastoralism increasingly difficult.

And yes, the rural population – exactly the people who are already dealing with the greatest problems – regularly faces another issue. The 6-tonne, long-nosed, great, grey hulk of an issue.

All along the edge of the Okavango Delta are small family farms, where local people grow crops to feed their families and sell in the market. One of the biggest complaints I heard when I was there was that elephants would come at night and eat whatever they had planted. I met one man called Jameson, whose fields had been raided a week before.

'They come when there is no moon,' he said, 'when they know they cannot be seen, and eat everything. They are always the big bulls.'

Jameson's story was a familiar one. I heard another man complain that six elephants rampaged through a three-acre farm and ate over half the crops in a single night. It's true that elephants tend to crop raid at night, and more so during a new moon, when light levels are lower and the elephants are less likely to be detected.

I met husband-and-wife team Graham McCulloch and Anna Songhurst, who set up the Ecoexist project in the 'panhandle' of the Okavango Delta, where 16,000 people live alongside and have to share resources with 18,000 elephants – the largest population of elephants in Africa living *outside* a protected area.

Graham explained, 'In this area, farmers plant crops in the fertile lands next to the Okavango River, usually harvesting them between January and April. And this period is also when crop raiding by elephants is most commonly observed – at the peak of the harvest. The elephants do it partly as a means of getting extra minerals like phosphorus, which they can't forage for in this area.'

Annual peaks in crop raiding do vary across elephant range, depending on local rainfall, but they typically coincide with periods of crop harvesting. In Gorongosa National Park in Mozambique, for example, peak crop raiding occurs in the dry season, between April and October, after the bountiful wet-season foraging has come to an end within Gorongosa, but the availability of mature crops outside the park is high. Most raiding occurs a few weeks before crops would have been harvested. And here, elephants seem to favour high-energy crops such as maize.

Graham continued, 'So, we have a pretty good idea when crop raiding is most likely to happen, which elephants are most likely to crop raid, and sometimes we even understand why elephants choose to crop raid. And now, we have increasing evidence of *where* crop raiding will occur. The elephants are not stupid, they tend to try their luck in isolated fields near to wild areas, rather than more densely populated areas. If there's a path, then it's easier for them to come in quickly and escape. We've seen the data from collared elephants, they have been tracked as running in and out of fields, as if they know that it's naughty and dangerous.

'The effects of an elephant can devastate a village if a farmer loses all his crops, but it's not just the economic loss that we have to worry about, it's had a profound impact on village life, too.

People used to sit around the campfire and socialise, but where there is a greater perceived risk, then people won't go outside at night and hide inside their houses because they're scared.'

The fear, it seems, runs both ways.

Of course, these problems are not unique to Botswana, with rural subsistence-level farmers across sub-Saharan Africa suffering devastating losses of crops, livestock, water-infrastructure, and even their lives in extreme situations. But the situation in Botswana is interesting, because it is home to the single largest elephant population in the continent, at an estimated 120,000 individuals. And this population has remained pretty constant in size over the past two decades, not suffering the same massive declines as have been evident across much of the rest of Africa.

Recent studies of elephant density and the carrying capacities of the environment, which we looked at in the last chapter, took into account vegetation type, water availability and current poaching rates, and suggest that most elephant populations still fall well below the sizes that could be supported in those areas. The continental population is only a quarter of what it could be for the space that is available today, *if* those spaces were connected up and elephants could move between them.

By connecting even the limited areas that remain available for elephants, we could be supporting an elephant population today more like that in 1979. In fact, only Hwange, in Zimbabwe, is really exceeding the benchmark population level.

Certainly, some of Botswana's elephants have been moving about, shifting ranges as rainfall patterns and the location of water sources and food resources change, with associated declines in number in some areas and increases in others. However, the overall elephant population within the country has not really altered in absolute number and despite the continued vociferous claims that the country is overpopulated with elephants, they could in theory support more. But that has missed the point.

Whether they are overpopulated or not, the increases in elephant density we see in these regions are a signal that disturbances are intensifying as a result of increased human–wildlife interaction.

For example, the human population of Botswana has grown by about 600,000 in the last twenty years, which is, of course, more than the entire population of elephants currently in Africa (although that is a pretty low human growth rate compared to many African countries). And like everywhere else in Africa, the human population is going to keep on growing – with Botswana's population predicted to increase by another 800,000 people in the next two decades.

If there are more people living in Botswana's rural areas now than there were twenty years ago, it is no surprise that more people are reporting problems with elephants and a common utterance is heard of 'too many elephants'. A new president has recently taken the reins in Botswana, and he decided to deal with these complaints from the rural electorate by calling for a programme of culling, opening up ivory trading, and also reinstating trophy hunting, which had been banned since 2014.

The culling proposal seems to have been recanted, for now at least, following a huge international outcry. Meanwhile international ivory trading decisions are out of the hands of one president or country, and luckily there is little appetite globally to legalise the ivory trade again. But with trophy hunting, Botswana can make its own decisions. And so, trophy hunting will resume in 2020.

The government argues that with 'too many elephants', they can use hunting to reduce or control elephant numbers and lift people living alongside elephants out of poverty. As you might imagine, the arguments this has caused have been intense and sometimes downright nasty.

First up, it is important to note that trophy hunting itself is unlikely to result in any real tangible decline in human–elephant conflict (HEC). Even if some of the males shot as trophies are also those responsible for some crop raiding, we know from the Amboseli population that there is likely to be a large pool of occasional crop raiders, and eliminating one or two, even a handful of them, is therefore unlikely to have any major impact on the overall number of crop raiding and conflict events.

Similarly, shooting the 150 or so males that are licensed for 2020 is not going to have any impact on overall elephant numbers (although if hunting rates increase substantially, it could result in demographic and genetic changes to the population, as we have seen). But I doubt the Botswanan authorities really think that trophy hunting will reduce conflict or elephant numbers directly. Instead, what they hope it will do is raise funds for rural development and increase the goodwill of people living alongside wildlife. Both of which are understandable and laudable aims.

Nevertheless, given how controversial trophy hunting is, we do need to scrutinise the situation in a bit more detail. Has conflict really increased since the 2014 ban on hunting, and will hunting revenue genuinely enable rural development? Perhaps most importantly, are there any alternative, non–lethal, actions that could be more effective in reducing conflict and encouraging sustainable development? And is the claim at the root of all this actually sound – at a time when continental numbers are falling rapidly, *are* there too many elephants in Botswana?

Despite the vast situational and economic differences between African nations, answering these questions for Botswana could be a model for conflict alleviation and sustainable development right across Africa, so they are worth thinking about carefully.

Everywhere I went in Botswana, people told me that they were encountering more elephants than ever before. Even my brave Bushman guide Kane said that he was fearful for his children sometimes when they went to school. But the matter of whether or not there really is more conflict is up for debate. It is certainly true that people are encountering elephants more now in Botswana – seeing them nearby more often – as the areas being used by both elephants and people shift. Elephants are currently moving back into areas of Botswana where they have not been seen for decades, and some of these regions are much more heavily populated and developed than they were.

But looking at the incidences of HEC reported within the areas where elephants were living up to 2014 – when elephant trophy hunting was banned – Mike Chase of Elephants Without Borders reports that conflict rates have remained constant. That is, the ban itself has not altered rates of conflict, so reinstating trophy hunting is unlikely to impact the actual number of incidences.

Human–elephant conflict is certainly a problem, but reports that it is intensifying could in part be due to perception. Droughts and shifting surface water availability make all aspects of rural life harder, and when things are difficult, we all tend to have a focus point where we channel our anger. Perhaps elephants fulfil such a role for some rural Botswanans. On top of that there is often a mismatch between the reality of crop raiding and people's perception of it. Large animals – elephants and also, for example, chimpanzees at the forest edges of Uganda – are blamed for much of the destruction, when other pests such as baboons, rats, bush pigs, or birds, are often taking more of the crops.

Importantly, the increased use of mobile telephones and social media has also meant that reporting of elephants spreads quicker, which can lead to rumours circulating faster than ever before. Compensation schemes are another problem. These efforts to offer financial reparation to farmers who have lost crops, or livestock, can end up confusing things.

These schemes are supposed to reduce anger, prevent retaliation, and encourage tolerance among individuals whose livelihoods have been damaged by animals. Damage must be reported to wildlife officials running the schemes, who then assess and quantify the extent of the damage and pay out accordingly. But hidden costs to farmers – economic and social – are not taken into account, only the market value of the lost crop or livestock animal. Such schemes are not always successful, therefore, in genuinely increasing tolerance, and may even increase resentment.

Compensation schemes can also make things worse by implicitly encouraging a 'human *versus* wildlife' perspective, and suggesting that interactions between wildlife and farmers are always costly. By pitting humans against the wildlife that they live alongside, and implying that the wildlife is at fault, compensation schemes may encourage antagonism and fail to inspire farmers to see themselves and the surrounding wildlife as part of the same system, where they could coexist.

This is not to diminish the perspective of Botswana's rural population: whether or not HEC has actually increased, the problem is serious and people *feel* like it is increasing, and it is therefore a problem that has to be dealt with.

I'll never forget a conversation I had with Kane in Botswana, when I asked what he thought was the solution.

'Lev, you know what,' he said, 'maybe all the countries in Europe and the rest of the world should take a herd of elephants each and let them roam free, in the way that you want us to have them to roam free, and see how long you last ... let's see what solutions you come up with.'

Although it was clearly a joke, he makes a valid point.

There are various strategies that individual farmers can adopt to alleviate conflict and discourage crop raiding, as we shall explore in the final chapter, but at a countrywide level, there are two main approaches to reduce conflict. The first is to control elephant numbers, the second is to change the perspective of people living alongside wildlife.

The Botswanan government is hoping that hunting will change perspectives and increase tolerance, by providing real financial gain to communities in the form of hunting revenue. There is usually also the gain of meat from the hunts after the trophies have been removed; the importance of this protein for many rural people should not be underestimated. Interestingly, at the same time, the government is going to stop compensation schemes for HEC, perhaps for the reasons outlined above, but it could also have the effect of focusing people's minds on hunting as a source of recompense for living alongside elephants.

It is true that income and employment for some communities next to hunting concessions fell away after the 2014 hunting ban. But there's an argument that rescinding the ban could damage Botswana's 'brand' reputation, putting its high-end and very valuable photo-tourism industry at risk – only time will tell.

Accurate data comparing income and employment generated by photographic tourists and trophy hunters is difficult to come by, but on balance, it seems that reinstating hunting may well offer some financial benefits to some communities in some areas but, as ever with trophy hunting, it may also be damaging in other ways. And remember, there are shamefully few examples of communities *actually* being lifted out of poverty by the income and employment they receive from trophy hunting concessions (although the same argument could be said of conservation tourism in general).

If trophy hunting is unlikely to alleviate HEC – either directly, by removing raiders or reducing populations, or indirectly by increasing the tolerance of people living alongside the elephants – are there any alternative tactics? Even if trophy hunting does end up having a positive effect, it is still good to consider alternatives, given how unpalatable an industry it is to many people.

Let's begin by thinking about the people, rather than elephants, and how we might increase tolerance and sustainable economic development in order to change or reduce perceptions of conflict.

'Fortress conservation' – keeping local people out of parks and reserves – has encouraged the view that the ruling elites and foreign visitors value wildlife more than the lives of the people living alongside elephants, which could understandably breed resentment. And let's be honest, in many ways that is true. Communities have been marginalised and pushed off productive land to form and maintain wildlife refuges, whilst receiving little or nothing back.

Maasai are asked to not let their cattle graze or drink in many of Kenya and Tanzania's national parks, because tourists don't want to see cows. The Herero people in Namibia were pushed out of Etosha National Park in 1907, and their ancestors today still report feeling a sense of loss. And the San of

Botswana have been moved out of traditional lands even within our own lifetimes.

When I visited the Virunga National Park in the Democratic Republic of Congo, there is a very clear line in the form of an electric fence dividing community land and national park. Any local person found inside the park is liable to be arrested and sentenced to three years in prison, or if they are armed and threatening may be shot on sight.

Imagine the reaction if all the people living on your street were told to move, because the government wanted to plant a forest where your houses and schools and shops currently were, and when you reasonably asked where you'd be moved to and what you would be given in return, you were pointed to a bare patch of wasteland and handed a shovel and some bricks.

There may be some substantial social wrongs to make up for, but it is hoped that by showing these people that they are respected as citizens, and their lives existing alongside wildlife are valued, then resentment and perceptions of conflict will be reduced. Empowering communities to manage their own safety and livelihoods should enable a view of themselves as respected decision-makers, rather than victims.

Of course, people need incomes, as well as respect. But given the booming wildlife tourism trade, it really can't be impossible to reconcile the needs of communities with the aims of wildlife conservation. People need to receive real incentives for conserving and protecting the wildlife they live alongside, with tourism proceeds distributed equitably. And people need to understand *why* conserving wildlife is important.

Perhaps demonstrating why so many people do value wildlife would be a good place to start. I have been amazed by the number of local people I have met in Kenya and Botswana who have never seen a wild elephant. Visits to national parks or reserves are 'what white people do on holiday', and it simply doesn't occur to many citizens to go and view their own national heritage.

Even people living alongside wildlife at the boundaries of reserves have often never been invited into the reserves as tourists – they've never had the chance to sit in a dusty Land Rover and watch the animals simply being animals, without conflict or competition with people. Encouraging local residents into reserves, increasing their familiarity with elephants and other animals away from conflict situations, could go a long way towards improving the perceptions of wildlife.

To this end, SANParks in South Africa now runs an annual 'Know your National Parks' campaign, where for one week it offers free entry for citizens to many of its parks. The scheme has been growing annually, and it is allowing tens of thousands of people access to wildlife viewing experiences that might otherwise be beyond them. On a much smaller scale, various charitable organisations offer similar schemes, but rolling out such opportunities on a larger scale really could have a significant impact in reducing the 'us versus wildlife' mentality.

The aim, then, must be for sustainable and genuinely inclusive development, which fosters coexistence between communities and wildlife. The UN 'Sustainable Development Goals' outline seventeen global aims for economic and social development around the globe, stipulating that each is achieved in a sustainable way without further compromising the well-being of the planet. They include goals such as 'no poverty', 'zero hunger' and 'good health'. Things it would be hard to deny are important.

The catchily titled 'Intergovernmental Science-Policy Platform on Biodiversity and Ecosystem Services' (also known as IPBES, thankfully) has recognised that nature and wildlife are essential for achieving the Sustainable Development Goals. But they also state that current trends in increasing habitat loss and extinctions could render many of the goals – including the 'big ones' relating to poverty, hunger, and health – unachievable. On a more positive note, they also state that nature is declining less rapidly in regions managed by indigenous people. Allowing local communities to take on more management decisions and empowering them to have a say about wildlife really could save the world.

Changing perceptions of people may not be enough, though – especially not in a country like Botswana where there are, undoubtedly, a lot of elephants. A lot is not the same thing as 'too many', but we will get to that.

Delineating parks and reserves – by fences, or with other boundaries such as settlements, agricultural lands and infrastructure – can restrict the movement and natural dispersal of wildlife in general and elephants in particular. And when elephants are confined, even in large areas, they do visibly change the habitat, reducing tree cover and diversity, for example, as they pull down and debark trees to feed upon. Culling emerged as the management tool of choice for many elephant range states – particularly in southern Africa – to prevent such 'damage' by keeping elephant numbers within ranges that people decided were acceptable.

In Kruger National Park, management decided that the land could not reasonably sustain more than 7,000 elephants, because the damage being done to vegetation in certain areas by elephants was deemed extensive. This was based upon the findings of a vegetation study conducted in the mid-1960s, which concluded: 'When the point is reached when elephant biomass for the entire park reaches the generally accepted carrying capacity for elephant (one elephant per square mile), the utilisation rate [of woody species] will have reached dangerous proportions.'

However, these inferences were based on the assumption that the woody vegetation present in the 1960s provided an accurate representation of a 'natural savannah'. But historically low numbers of elephants (and other large herbivores), due to excessive hunting, likely resulted in much greater woody cover than would have been present prior to European settlement. Furthermore, the decision to implement population management was also made at a time when rainfall was below average for five successive years. From 1967 up to 1994, 14,629 elephants were euphemistically 'removed' from Kruger.

Culling was suspended in 1994 whilst a comprehensive review was conducted, and essentially banned in 2006, on the grounds that other management options would be more effective. You see, every time after a cull had taken place, the growth rates of the population shot up to around 6.5 per cent (very close to the maximal rate), with intervals between calf births often going as low as three years for many females. People would kill elephants, opening up space in the ecosystem for more elephants, so the elephants would respond by having more calves.

Eventually, policymakers heard the shouts from various scientists, and decided to listen to their explanations of why this baby boom kept happening. Being located in a naturally dry area and yet attracting a lot of tourists who wanted to view wildlife, the management of Kruger had, over the years, opened a number of artificial water sources. These provided excellent viewing opportunities for visitors, and also meant that animals such as elephants, who need a lot of water, didn't have to move very far in the dry season as they otherwise would have – they could hang around these artificial water points drinking to their heart's content.

By discouraging normal elephant movement and concentrating a lot of elephants in these water-provisioned areas, the vegetation around these water points – within a mile or two radius – was being completely hammered. But the elephants were happy, as they had enough food and water to breed at high rates.

Policymakers eventually decided that instead of direct management of numbers – culling – managing access to resources could be a better option. So, many of these artificial water points have been closed since 2006, and elephant birth rates have slowed. The total number of elephants in the park has increased without the high death rates – now standing at around 18,000 individuals – but the elephants are more spread out over the entire park area, reducing the pressure on vegetation in any one space.

Whilst debates about culling continue to rumble on in southern Africa, with some trigger-happy 'diehards' (excuse the pun) still claiming it is necessary, and the best solution to reduce the 'too many elephants' they keep seeing, all the scientific evidence points towards natural resource management being a better long-term strategy. Elephants do have an impact on tree structure and abundance in Kruger and other large protected areas, but this does not necessarily reduce habitat quality for other wildlife, so there is no need to try to lower elephant numbers artificially. And science seems to be winning out with policymakers, on this issue at least, with culling becoming rarer and rarer.

Moreover, natural resource management avoids the very unattractive side effects of culling, such as having to kill entire families, including calves, or – to some people's minds the worse option – intentionally raising juvenile elephants as orphans. You'll remember that the crazily hormonal adolescent males of Pilanesberg were Kruger-cull orphans. And the female orphans did not come out of this unscathed. Recent experiments demonstrated that the social trauma experienced by young females who witnessed the culling of adult family members resulted in greatly compromised social functioning decades later.

Unlike the 'natural' family groups in Amboseli, which have not experienced such profound disruption, the adult female elephants in Pilanesberg – all of whom were Kruger-cull orphans – failed to distinguish other family groups appropriately and to assess the level of social threat that they presented. Their social knowledge and functioning were disrupted into adulthood, which could have major impacts on their behaviour and calf-rearing strategies. In fact, some people liken the effects of elephant calves witnessing the culls of their mothers and aunts to post-traumatic stress disorder: the effects of the early trauma certainly do seem to persist long-term.

Managing water points in Kruger has been deemed a success, because it has changed the space use and density of elephants across the landscape. But ultimately, the elephants are still constricted within a park. Admittedly a vast one, but villages, towns, roads, railway lines and fences still restrict and define where the elephants can move. Across many elephant range states, fencing is about the most common and obvious method to mitigate conflict – keeping elephants inside a reserve, and people out. But these fences also isolate elephant populations, curtailing their connectivity, and possibly leading to degradation of the habitat – as is the case in many of South Africa's smaller fenced reserves.

Also, as genuinely 'elephant-proof' fences are pretty hard to come by, breakouts can occur that damage infrastructure and further contribute to the conflict narrative. Even in the absence of fences, elephants seem to know where protected areas end, and so where they become more vulnerable. Data from GPS tracking collars on elephants across the continent shows that, currently, elephants often streak through these unsafe areas as fast as they can, in the search for more secure habitats. Analysis of stress hormones in their dung tells us they are very stressed as they do it.

The best solution to all of this – in fact, the best solution to almost all 'elephant problems' – is the creation of 'mega-parks'. Essentially, this means connecting up the remaining elephant habitat via the use of corridors and forming 'transfrontier conservation agreements', allowing the historic connections and dispersal patterns of elephants to become re-established.

We know that as the number of humans increases, elephants want to move away to areas where the human density is lower. However, moving away requires both the land to move along, and land to move to. At present, many elephant pathways between feeding areas or neighbouring populations are themselves blocked, being populated with humans or our infrastructure, and so conflict ensues.

Often, these blockages occur quite innocently, simply because the people moving in did not realise the land was important for wildlife movements: it can require years of animal tracking data to identify areas as corridors – literally, strips of land of varying width, depending on the country and habitat – that allow animals to move between key habitats and ranges. But even if the intent was not malicious, the effect is equally damaging.

The Amboseli ecosystem stretches far beyond the national park boundaries, across an international border into Tanzania, as far north-west as Nairobi National Park, and east to the Chyulu hills and beyond. But expanding agriculture and human settlements have significantly narrowed many of the corridors that were previously used by Amboseli elephants. Of eight identified corridors, two have been blocked entirely, three are in danger of being blocked, and only one is formally protected, making the ecosystem and elephant population vulnerable.

Identifying, reinstating, and preserving corridors will have significant impacts. Corridors can allow elephants to move to new areas to find water and food, and so reduce the feeding pressure on any one area. They can allow males to find new females to mate with, and so maintain genetic diversity and viability. And, crucially, they can allow elephants to move away from humans – and so reduce instances of conflict. The corridors themselves may not offer much in the way of food or water, but they offer highways to other areas of good habitat.

Opening up corridors and connecting 'mega-parks' does not mean removing all fences. Many fences will be necessary. For example, in Botswana, beef exports make a significant contribution to the economy, and beef cattle have to be kept separate from wildlife for import into Europe, to prevent the spread of diseases. But better planning of where fences and boundaries lie, and where elephants can and cannot move, could go a long way to alleviating the pressures of HEC.

The location of corridors is key. If they run past lots of smallholder farms, elephant raiding will continue. It is even possible that some people will need to be relocated. Not in the style of old, but sensitively, voluntarily, and with appropriate compensation and incentives offered – such as funding to build better housing and new schools.

Assuming that collaborations between all affected parties can be achieved, corridors and mega-parks really do seem like the only way to go. Allowing elephants to disperse and move naturally, recolonising areas that they currently cannot access, whilst at the same time allowing the people they live alongside to be involved in the decision-making and funding processes – this can all help address the central causes of human–elephant conflict.

If elephants can disperse, we will soon see a reduction in density in the areas that are currently highly populated. The elephants in Botswana are increasingly moving into areas more densely populated with humans, not because they like people, but because they need to move to areas with adequate food and water. Opening corridors to areas with similar food and water but fewer people would be in everyone's interest. It may sound fanciful, but such areas do still exist, and the southern African range states are genuinely trying to achieve such mega-parks and safe corridors.

Botswana is working closely with Angola, Namibia, Zambia, and Zimbabwe to enhance the 'Kavango Zambezi Transfrontier Conservation Area', or KAZA – which across its extent is home to more than half of the continent's remaining elephants. The hope is that improvements across the KAZA area will pull elephants out of the densely populated areas of Botswana and Zimbabwe, into suitable habitat in Angola and Zambia. Remember, many of the elephants in Botswana probably originated from Angola, moving to escape the civil war and high poaching rates.

If habitat in Angola can be secured, and safe corridors provided, I think the majority of elephant scientists would agree that the elephants will move back. The movement may take some time, but it will be a long-term solution, reducing all the pressures on the areas that currently feel they have too many elephants. *If* safe passage and protection can be assured.

There is the strange tale of the elephants of South Sudan. In 2011, after thirty years of civil war, elephant populations had been virtually decimated due to poaching and hunting to sustain the various armies that roved the plains and swamps. Locals said that the elephants had 'gone south' into Uganda or DRC. Then, as soon as the fighting ended, within ten days there were reports that elephants had crossed back into the country. Quite how they 'knew' the war had ended, nobody is sure, but somehow the first curious bull elephants had gone exploring, reported back, and now some breeding herds have returned – at least temporarily.

So as long as we don't allow killing rates to spiral out of control and learn appropriate ways to share land, elephant populations still have a good chance of rebounding.

A transfrontier conservation area is also being managed across South Africa, Zimbabwe and Mozambique: the Greater Limpopo TFCA. This initiative has removed international boundary fences that used to separate Kruger National Park from Gonarezhou National Park in Zimbabwe and Limpopo National Park in Mozambique, and now incorporates many additional adjacent protected areas, allowing inter-connectivity and movement between large but formerly isolated elephant populations.

Even if conflict with humans was not such a pressing problem, the creation of such mega-parks and corridors would still be vitally important for the continued survival of savannah elephants – as a means of buffering against the effects of climate change. Southern Africa is home to the largest populations of savannah elephants, but these southern ranges are the most vulnerable to future droughts, with the climate in many areas already becoming drier.

This will place greater strain on the elephant populations, vegetation, and water resources in these areas. And if the elephants cannot disperse widely and seek out new water sources, then destruction and death to elephants, plants, and people could spiral out of control.

Scaling up this trend in creating mega-parks that link and reconnect elephant habitats means that the high elephant numbers we discussed earlier as being *possible*, could become a reality – and without necessarily increasing conflicts. In essence, elephants want to eat, drink, and avoid people. Connecting the habitat areas can allow them to do that, as long as the reserves and corridors are well managed – with conservation and sustainable economic development aims given equal precedence.

Perhaps, too, we should stop using the word conflict, and instead talk about human–elephant overlap; after all it's their land too, and it's hardly a fair fight.

11

A World Without

As the sun was setting across the Okavango Delta, I looked out across one of the last great pristine wilderness areas left in Africa. In amongst the tall papyrus islands, herds of elephants gathered to congregate in the lush swamp, where food is abundant and water plentiful. I found myself speechless at the sight of a mother and calf swimming across a river with their trunks raised like snorkels, while overlooking hippos grunted in unison. It was truly a sight to behold, especially as the matriarch turned to help shove her baby up the bank in an act of motherly love.

But it was also a sight tinged with sadness, as I knew that it was unlikely to be this way for much longer. Every year we build more roads and fences and in doing so encroach upon more and more wilderness, depriving elephants of their ancestral migratory routes. As the human population grows, in equal measure so does the likelihood of elephants becoming extinct, even within a generation.

It feels like we have reached a critical junction. We can act now to conserve the remaining elephants, or we can continue to concentrate on our own greedy needs and let elephants meet the same fate as their mammoth cousins.

Why should you care? The more cynical reader might say that it doesn't really matter either way if elephants survive or not. We humans have done alright without mammoths for the last 5,000 years, so why worry too much? I think it goes without saying that I'd be upset if my future children and grandchildren never got to see an elephant in the wild, but in the same vein, I'm sure a lot of African farmers wouldn't lament their loss too much. As my guide Kane said in Botswana, if we all took some elephants home and let them roam free in our gardens, how long would *we* last?

It's a fair point. Elephants were wiped out in Europe a long time ago, because of hunting, human greed and climate change. What's different now? Well, given that we are indeed facing that loss imminently, and this time round we actually know what we are doing, it is worth considering what a world without elephants would really be like.

Elephants are not the only species threatened with extinction, of course, and certainly not the only species under threat because of our actions. The rate of change and degradation to natural landscapes, habitats, and wildlife in the past fifty years is unprecedented, and we humans are driving that degradation. Given how close so many species are to extinction, you won't be surprised to hear that a lot of people have been considering exactly the kind of question we are asking – how any one species or habitat benefits us and what effects their loss might have on people around the world, even if we think we are far removed from nature.

People often now use the term 'natural capital' in this vein, because as a species we seem compelled to bring things back to money and economics. It is really a shorthand way of asking what are the costs and benefits of nature? What rewards does nature bring us, overall?

Understanding these costs is important because, as the Namibian Minister of Environment and Tourism, Hon. Pohamba

Shifeta, said recently: 'Just as a private sector investor will not invest in something without knowing its likely returns, the Government must also know the value of nature, who is benefiting from it as well as the type of returns it is generating. This is vital to inform our planning and budgeting processes.'

His words may sound vaguely disturbing – that governments will not invest in nature unless they see the value – but that is the harsh reality. And the thing is, once we start quantifying the value of nature, and policymakers and society really understand what it does and what it provides, I'd like to think there will be a massive upswing in funding for conservation. Because we fundamentally need nature and nature needs elephants.

Nature is not a waste of space that could be better used for housing or farmland or industry. Wild spaces, and the animals they support, provide us with a huge range of fundamental 'services' – ecosystem services, to use the fashionable terminology – that confer real benefits and values to all people.

Ecosystem services include such fundamental things as providing, storing, and purifying water; capturing and storing carbon; and cycling vital nutrients; as well as maybe less obvious things like buffering extremes of drought and flood. The increasing trend in the UK to revert to natural flood defences is because we are remembering that nature – when allowed to work properly – is pretty good at this kind of thing. When we damage nature or natural processes, we lose these ecosystem services, which undoubtedly makes us worse off financially and socially.

Perhaps we even lose out from a well-being perspective, because ecosystem services also include less tangible 'cultural services'. The recreational, historic, and aesthetic benefits that we ascribe to individual species or wild places are important factors in their value, even if we can't put a price tag on them. So just as man-made structures like the Notre Dame Cathedral in Paris or the Great Wall of China may be described as having 'huge cultural significance', so too might a natural place like Northern Ireland's Giant's Causeway, or an iconic species – such as elephants.

To better understand what the world would be like without elephants – and make a convincing case for conserving them – we can employ this natural capital framework to consider what benefits elephants provide us. Exactly what services do elephants deliver, and how do these measure against the financial and social costs of conserving them?

Elephants contribute unique services that benefit us. Their ecological impacts can be extensive, and directly and indirectly support a range of ecosystem services. They have been characterised as 'ecosystem engineers' and 'mega-gardeners', based on their role in dispersing seeds and changing the appearance of landscapes, making elephants a key force in maintaining the health and structure of savannah and forest ecosystems. Plus, they offer a range of cultural services.

By eating fruit from trees and then distributing the hard kernels in their dung as they walk, elephants spread seeds far and wide. In the tropical forests of Thailand, Asian elephants are the most effective seed dispersers around, out-performing gibbons, deer and bears in the role. And in central Africa, forest elephants are known to move great quantities of large seeds away from the parent tree, and so maintain forest structure and diversity. Even savannah elephants have an important role in seed dispersal, distributing marula, baobab, mongongo, and various palm fruits over distances of many kilometres.

The loss of elephants in important tropical forests of Africa is going to have dire consequences for the future survival of some tree species, which will have knock-on effects for carbon capture and storage, and therefore climate change. Currently, the African and Asian tropical forests, together with those of South America, absorb nearly a fifth of all the carbon dioxide released by burning fossil fuels.

Without elephants to spread viable seed pods, fewer seeds will germinate and grow into saplings, and so fewer trees will grow to replace the older ones that die or are felled by people, and the role of these forests as significant carbon stores will be greatly impacted. Using carbon offset prices, it has been calculated that elephants are performing a carbon capture service in encouraging the new growth of trees that could be worth millions – even billions – of dollars.

Elephants also change woodland and forest structure by pushing over, uprooting, and debarking large trees. By reducing the density of old trees, which may sound bad in itself, they create space for new trees to grow, improving the amount and quality of food available for smaller browsers – the herbivorous mammals such as bushbuck, kudu, and black rhinos that mostly eat leaves, bark, and greener branches from trees and woody plants.

Elephants can open up pathways through thick, spikey areas that other animals can then use to access plant foods. They can even convert new woodland areas, which grew up in the past few hundred years as elephant densities declined due to heavy hunting, back to their previous savannah grassland states, reopening historic environments and feeding opportunities for a whole host of animals.

In fact, the herds of impala and gazelles that are synonymous with our views of 'Africa' often do better living in environments with elephants, who maintain the grasslands these antelopes need. Elephants can even help cattle access more food, by increasing the quantity and quality of their diet during the wet season. Yet more evidence that the narrative of competition and conflict between elephants and farmers can be misleading.

Elephants don't only engage in forestry services. They are also excellent at more general landscape design, creating and maintaining pathways and water features that deserve a stand at the Chelsea Flower Show. Elephants create trails and pathways through habitats that many other creatures use for ease of travel. In central African forests they maintain clearings, the water and mineral-rich soils of which become important mineral sources and gathering points for many forest species, including gorillas, buffalo, and forest antelopes.

They remove sediment from standing water when they mud-wallow, keeping waterholes open, and often open up new ponds or sources of water. On top of that, the 50–150 kg of dung they may produce per day (yes, what they can poop out per day can be the equivalent of two fully grown men!) returns nutrients to the soil over a vast area. As if that wasn't enough, elephants are also excellent fire marshals. Through their bulk foraging approach, they reduce the amount of dry grass and leaf litter that can accumulate around trees, and so reduce the risk of bush fires burning out of control.

Poaching and conflict with livestock herders has driven elephants out of many areas in northern Kenya. As the elephants disappeared, the herders noticed that so too did natural waterholes and paths through the scrub land. Livestock were no longer able to penetrate the vegetation to feed on mountainsides, so herders began to set fires to clear the bush, eventually resulting in deforestation, soil erosion and drying of water sources. As elephants have been encouraged to return to the area, waterways and pathways have reopened, and damaging man-made fires are no longer necessary.

Elephants don't provide services purely for people and other mammals. They are also pretty useful for many small creatures, too. In some forest environments, the footprints that elephants leave in mud have been shown to be important for frogs, because the footprints can fill with water, providing pools for frogs to lay their spawn. These miniature pools can also act as 'stepping-stones' for frogs to move between larger pools, connecting up different frog habitats.

Let's not forget the geckos, who prefer to live in trees that have been damaged by feeding elephants. The bark stripping and branch breaking of elephants provides a lot of hiding places for the little lizards, making these 'damaged' trees much more safe and attractive places to live. Even elephant dung provides an important home and food source for many different kinds of creepy-crawly, from dung beetles to scorpions and millipedes. And larger animals such as meerkats can often be seen rummaging through piles of elephant dung, looking for these bugs to snack on.

It is clear, then, that elephants fully deserve and should take pride in their labels of 'ecosystem engineers' and 'mega-garden-ers'. They don't passively contribute to ecosystem services; they actively promote and upgrade such services. However, these benefits only play out when elephants are not squashed into restricted areas without corridors allowing them to move between key habitats. Elephants confined at high densities can be very bad news for local plant and tree life and, therefore, for other animals that depend on those trees and plants.

When we mess up the balance for elephants, we mess up a whole range of critically important natural services that we rely on. Elephants need space – corridors and mega-parks – and it would seem giving them this space is entirely in our best interests.

Elephants do have other value, beyond these ecosystem services. Even elephant dung has other uses. It can be turned into a source of biogas; used as a mosquito repellent (the smoke from burning it, that is, rather than having to smear it on directly); and even to make paper. You may have seen on popular survival TV shows that if you are really desperate, squeezing water out of fresh elephant dung can be a life-saving thirst quencher, although I'd suggest you leave that as a last resort.

Perhaps more importantly, elephants are icons of Africa, attracting millions of visitors every year to the many African countries that depend on tourism revenue. Eco-tourism generates obvious economic incentives for nature conservation, and it has been shown repeatedly that elephants are one of the most popular species, especially amongst first-time and overseas tourists. The allure and appeal of elephants make them a 'flagship' species, attracting revenue and interest. This revenue can then be used to propel communities living alongside the elephants out of poverty, and further conserve nature.

The tourism revenue that has been lost to African countries as a direct result of losing elephants in the poaching crisis has been calculated at $25 million per year (since the poaching resurgence); a price that exceeds the current costs of anti-poaching enforcement! Protecting elephants makes financial sense.

Throughout our shared history, elephants have provided us with these vital ecosystem services, and have provided a substantial source of income – initially through their tusks, and now through tourism as well. They have also provided us with meat, a military force, and a workforce, to varying extents. There is no denying that elephants always have been, and will hopefully continue to be, extremely important in a practical sense to our lives. We may not always be aware of it, given the complex nature of many of these ecosystem services, but without them our world would be degraded considerably.

But that is not all.

This emphasis on economics – on the services and benefits that elephants and nature provide – is a bit distasteful for many people. The 'commodification' of nature can feel wrong, however much we realise that knowing the explicit value can incentivise governments and agencies to invest time, thought and resources in conservation. We know realistically that nature does have to pay its way, but we don't want to think about price tags. I believe part of the reason that many of us struggle with this is because we are aware that nature – and elephants in particular – offers so much more than economic services. We want elephants to be conserved because they are magnificent, and mysterious, and wild. Not simply because they bring in the dollars.

Elephants do seem to be special, to attract a certain awe among people throughout the world. The Mali Elephant Project, a 'grass-roots' conservation organisation that works to protect the remnant population of desert-adapted elephants in the Gourma region of Mali, surveyed local communities as they were beginning to set up their operations, to try to get a handle on exactly what local people felt about the elephants.

This was at a time of severe conflict between people and the elephants, with intense competition over access to water and dwindling local resources. Yet despite this competition, nearly a fifth of the people asked stated that elephants were intrinsically valuable, simply for being elephants. They viewed elephants as lucky! And they were viewed as an important part of the Malian national heritage.

In the 'Know Your Parks' scheme that is run annually in South Africa, which allows free access to national parks for South African citizens, the park that receives the most visitors by far is Addo Elephant National Park (Kruger is not part of the scheme). Surely the name tells us something about what people are hoping to see.

At the very least, it tells us that people like seeing elephants for free. But what do I mean about elephants being 'intrinsically valuable'? In essence, it means that species are valuable irrespective of what they provide or take away from us. By virtue of being here, they have a right to exist, and it would be wrong for us to do anything that encourages their extinction. It's a perspective shared by many people today, from David Attenborough to Pope Francis (despite running against some entrenched religious perspectives, which inherently claim that people take precedence over animals, and the only value of animals is in what they can do for us.) I guess it is a personal call, but it may be one that is worth all of us dedicating some time to thinking about.

Whatever you think of elephants' intrinsic worth, they undeniably hold significant *cultural* value for us. Depending on where we are from in the world, elephants are religious icons, symbols of national heritage, of strength, wisdom or family. We tell stories about elephants, and myths and legends abound. And we use their image in a whole host of ways, in art, in books and illustrations, on logos. The list goes on. We like seeing and talking about elephants.

Elephants are known in the conservation world as 'charismatic megafauna' – large, attractive animals that get a lot of attention. You know, things like pandas, polar bears, gorillas, tigers; the animals that grace the cover of all the glossy conservation campaigns. And elephants are perhaps one of the most charismatic of all.

Some conservationists can be a bit sniffy about the 'value' of these charismatic animals. Understandably so – the plight of elephants does indeed attract a lot more attention from the general public than some critically endangered warty toad or hairy spider. But it may be fundamentally bad to be a charismatic species, because the general public finds it so hard to believe that such iconic animals – whose images are ubiquitous – could really be in danger.

In a recent survey, hundreds of people were asked what they consider to be the most charismatic species (elephants were third in the resulting list, behind tigers and lions). The ten species that were reported as being most charismatic by those asked are all endangered, but these same members of the general public answering the survey did not realise the animals were in trouble. Incredible. So, whilst some people might be bored of seeing elephants and lions and the like paraded around as the poster children for conservation, the message is evidently not getting through to everyone.

It is muddied, perhaps, by the fact that stylised images of these animals appear on so many logos, cartoons and adverts; we assume that if there are so many on our screens and in our shops and homes, there must be as many in the wild, too. 'Pandas can't be endangered. I saw hundreds of cuddly ones in Hamleys toy shop last week.' That sort of thing.

The very appeal of elephants, which we often assume is key to conserving them, could be promoting ignorance of their plight. Indeed, scientists studied some of the most popular documentaries of recent years and noted that while threats to nature are being mentioned more than in the past, there remains very limited footage of these impacts. Ultimately, this may result in failing to communicate fully the severity of the threats facing the natural world, particularly as the visuals focus on presenting a pristine view of nature that is separate from humans.

Yet the appeal of elephants is very definitely promoting their cause in other ways. Consider the 'Non-Human Rights Project', for example. It is not some kind of evil, anti-Amnesty International organisation trying to return us to the Dark Ages, but is an organisation trying to change the law in the US to give something equivalent to human rights to non-human species. (The hyphen is very important in their name.)

The legal mechanics are dense – all about autonomy and 'personhood' and a whole lot of other terms that don't seem to mean quite the same thing to lawyers as they do to you and me; but the essence is that this group wants social, large-brained, and intelligent species to be given particular rights so that they no longer have to be confined in captivity. They are fighting this ethical issue without making it about welfare, because typically that has not got animals very far in judicial systems.

So far, they have brought cases on behalf of four captive chimpanzees and three captive elephants, living in America. The cases are in various stages of appeals, with one of the elephant cases literally being laughed out of court at the first hearing. But they keep chipping away at the arguments, and they are making more and more people think that just because we are human, do we have a right to confine and remove the liberty of other social beings?

Before you dismiss it as doomed to fail, you should know that a court in Argentina awarded human rights to a captive orangutan in 2014. The first victory of its kind, this orangutan, who had been kept alone in a zoo for twenty years, will live out the rest of her days at a sanctuary in Florida, as the first official non-human person. Even great-ape people retire to the Sunshine State.

Whether or not you think that elephants should be given rights equivalent to our own, it feels pretty clear to me that we often take elephants for granted, not always giving them the respect they deserve (in the wild or in captivity). But it is also evident that we need elephants, in a host of different ways, for our physical well-being, given their role in maintaining natural ecosystem services; and for some kind of mental or spiritual well-being. It is not crass to say that elephants have value. In fact, we should be shouting from the roof tops about how much we need them and what they do for us, to make policy-makers listen.

Investment in conserving elephants *will* give good returns! Vital returns. Nature – and elephants within nature – are essential for our existence and quality of life. And by 'our', I do mean all people, everywhere around the world.

There's one last thing I want to touch on here: it's the story of the loneliest elephant in the world.

Knysna is a small town surrounded by the Tsitsikamma forest, on the southern coast of South Africa, some way to the west of

Addo Elephant National Park. Large herds of elephants used to roam here, sharing the area with the indigenous Khoesan people. As in nearby Addo, the arrival of European hunters in the eighteenth century quickly resulted in the near total elimination of the elephants. But unlike in Addo, the elephants inhabiting this forest area were not given special protection, with the area only proclaimed as a national park in 1964. By this point, it was thought barely ten elephants remained, but ages and sexes were unknown.

Being isolated from other elephants – Addo is the nearest wild population, 300 kilometres and several major highways away, and behind a very big fence – the future of the Knysna elephants was uncertain. The population, though only a handful of elephants, was very secretive, hardly ever being seen, but they left signs – dung, broken trees, that sort of thing. They took on a kind of mythical, cult status within South Africa and around the world, with people debating how many were left, where they moved, whether or not they had magical powers of invisibility or tele-transportation.

Camera traps were set, genetic analysis of dung samples was conducted, and countless survey transects were walked. Occasionally someone would catch a glimpse, but no one knew for sure how many elephants were left. In an attempt to revive the population, some of the Kruger-cull orphans were introduced to Knysna in the early 1990s, but they never adapted to the forest habitat (which with hindsight seems obvious – having no experienced leader to show them the way, how could they have learnt to live in a totally different habitat), and were removed a few years later.

By the late 1990s, two camps were appearing – those who believed only one individual remained, and those who thought

– hoped – that it was a small family of four or five. DNA analysis from dung samples seemed to suggest five or more animals, but in the brief glimpses people did get, they could only ever be sure of seeing one. Recently, the jury came back with the verdict. Eighty camera traps were set at forty locations across the entire forest range, and were left on for fifteen months.

They only ever photographed one elephant. A single adult female of about forty-five years of age, who has been named Oupoot (big foot). Oupoot should be in her prime, leading a family of adoring daughters and granddaughters. Instead, she is entirely alone, a painful, tragic reminder of what we – with our lust for ivory and greed for land – have done to elephants.

If Oupoot doesn't convince you that a world without elephants would be a sorry thing indeed, nothing will.

12

Sharing the Future

The United Nations predicts that the global human population will soar to somewhere near eleven or twelve billion people by the end of the century. It's a terrifying thought if you consider the impact we are already having on the planet, and what extra pressure an additional three or four billion people will have.

The threats to elephants are real and immense in the face of such challenges, and it is easy to become despondent and give up hope. Across the world we continue to turn a blind eye to the many injustices facing wildlife and allow terrible crimes against nature to be committed. Often, we ourselves are to blame; we choose the easy option and spend our time expressing faux outrage at a picture we see on social media, instead of doing anything meaningful. We might follow a charitable page and give it a like on social media, maybe we even donate a couple of pounds to a cause, but we carry on committing the same old crimes and fail to address the root causation of the issue.

Even for those who do make a change for the better, it's hard not to feel like it's a futile effort when up against the intractable rise of 'development' and the output of billions of people who simply want to survive. We know full well that beef farming and soy plantations in South America are causing the biggest environmental degradation, and yet there is little slow-down in consumption. We continue to use plastics and waste food and refuse to talk about the unsustainability of the human growth in the developing world for fear of hypocrisy or accusations of cultural superiority.

Well, unless we do address the elephant in the room, it will be too late to conserve our last giants. The problem is a global one, and we must all play our part to ensure that everyone can effect positive change.

Thankfully it's not all doom and gloom when it comes to the numbers. The UN also predicts that when the inevitable growth reaches its peak in eighty years' time, there will be a global slow-down as more people emerge out of poverty and better wealth distribution is enabled. Ultimately, the only way to reverse the process of unsustainability is to improve education across the board, by supporting emerging markets to redress the inequality through empowerment and wealth creation. In a nutshell, we need to work with the developing world to help them progress to a point at which they care. Education is the key to a more considerate, responsible and less reckless world.

I'd go one step further to state that it is in female education particularly that the balance of power hangs. There are 65 million girls out of school and 80 per cent of these girls live in sub-Saharan Africa. In Nigeria, 5 million children aged six to eleven years old do not have access to primary education. And the average literacy rate in Nigeria among females is less than 50 per cent. Why? Well, there are practical factors such as distance to school, cost of education, transport, uniform and books, but also many girls are forced or obliged to stay at home to help with domestic work through religious and cultural pressure – simply because they are girls.

The Nigerian fertility rate is 5.7 births per woman, one of the highest in the world. Imagine, however, if the average Nigerian girl is encouraged and empowered to stay in school until she is eighteen, instead; by default she will reduce her number of children by at least one or two before the age of twenty-five. What's more, she will be more educated, and rather than stay at home producing yet more children, she will be more likely to get a job and be informed to make her own choices; to have a say in her own culture and go on to implement greater societal change across her country. Her children will do the same.

If we can do this across Africa, then in one generation great change can be achieved. So, if we are to focus our efforts on something, then we need to start investing in anything that keeps girls in school longer, and that encourages employment and hope. By this logic we also should invest in healthcare, which reduces the societal desire to pump out endless children.

Bill and Melinda Gates run a philanthropic organisation that has spent billions of dollars to save the lives of millions of children in extreme poverty by investing in primary healthcare and education. Yet they come under constant attack from well-meaning people who say, 'If you keep saving poor children, you'll kill the planet by causing overpopulation.'

Of course, there is an inherent moral imperative to save dying children, but more than that we need to do everything we can to reduce child mortality, not only as an act of humanity for living, suffering children, but also to benefit the whole world now and in the future. Why? Because as counterintuitive as it sounds, the more children that are saved now, the faster population growth will slow down in the long run. As billions of people are emerging out of absolute poverty, most of them are deciding to have fewer children of their own accord.

People in the 'developing world' have quickly realised that they no longer need such large families to provide child labour on farms. And as healthcare increases, they no longer need extra children as insurance against child mortality. As women and men get educated, they start to want better-educated and better-fed children, and the answer is not to have so many. Quality over quantity. This is helped greatly by the availability of modern contraceptives, which let millions of parents have fewer children without having less sex.

Voluntary family planning is one of the most cost-effective investments a country can make in its future. Every dollar spent on family planning can save governments up to six dollars, which can be then be spent on improving health, housing, water, sanitation, and other public services, and of course – conservation.

Only the poorest 10 per cent of the global population now have more than five children. The rest of us are choosing to have fewer and fewer, and in countries like Iran, Brazil and China, the average fertility rate is under two. That's below the UK. Even in Afghanistan, where in 1995 the average woman had eight children, now, thanks almost exclusively to female education, it is down to 4.5 children.

When looking at this critical juncture in our history and the biggest challenges facing the planet, I think we have a moral imperative to help raise the average levels of education across the board, which in turn has the knock-on effect of reducing human population growth and its unsustainable side effects.

If the elephants are to stand any chance at all, we need to find a way that sustains the means of wealth production, by bringing more people out of poverty in order to save the environment, and ultimately, ourselves.

It is what we do right now, though, that matters in the short term. We need to promote better coexistence between elephants and humans; and coexistence *is* more common than we often realise, with many people living peacefully alongside elephants much of the time, across Africa and Asia. So let's try and keep some hope alive.

There is a lot we can do to change things, from mitigating against crop raiding or damage on a local level, to opening up corridors connecting the areas where elephants live, to encouraging sustainable development for the people living alongside elephants, and putting pressure on the relevant policymakers to make sure ivory is off-limits. It all sounds so straightforward, doesn't it?

Of course, things are rarely that simple in practice. All conservation plans we make must take wider economic, social, and ecological contexts into account, and things would be easier if we could predict the future. But one thing *is* straightforward: we have to act now to find solutions that work for both elephants and people.

Finding a way for people and wildlife to share land and resources is the ultimate conservation challenge. And given that we know elephant movement patterns are related to human activities, our conservation strategies must take account of the different socioeconomic contexts found across elephant range states. Conservation successes will continue to depend on sustainable human development; improving education, literacy rates, and good governance, whilst reducing corruption and inequality, are all crucial for protecting the natural world.

The immediate, most urgent threat to elephants remains poaching for ivory, but habitat loss and fragmentation are the long-term issues that we simply cannot disregard. Whilst anti-poaching enforcement and ivory-demand reduction strategies are therefore critical, we don't have the luxury of ignoring the habitat problem and dealing with only one issue at a time.

We have to tackle all threats concurrently and holistically. Lifting the communities in human–elephant overlap areas out of poverty, and encouraging their development in a sustainable, environmentally aware manner, will likely reduce the temptation to poach. It should also reduce the constant background calls to reopen the damaging legal ivory trade, as the need to raise national funds for conservation can be met in other ways.

So, we know what we are aiming for. But how on earth do we get there? I mean, eradicating global corruption and inequality seems like a pretty big task to heap onto the shoulders of a few conservationists. Like all things, I suppose, we have to take it one step at a time – with a broad consortium of conservationists, development organisations, government agencies, and civil society all working together.

Let's assume, somewhat hopefully, but with a bit of evidence behind our faith, that anti-poaching and ivory-demand reduction strategies will continue, and that the ivory trade will wither and crumble. It may not happen entirely, but we need demand for ivory to drop to the point where elephant poaching falls dramatically. Permanently. There is a lot we can do at the same time to counteract local-scale conflict and wider habitat loss.

Achieving coexistence is never going to be as easy as drawing a line between humans and elephants: we do not and cannot live separately. There is no one solution that will eradicate conflict, but with a lot of patient research and trial-and-error learning, we do now have a range of measures that can be combined and adapted to discourage or prevent clashes between humans and elephants.

Crop raiding by elephants is not a new phenomenon, having been recorded as soon as Europeans brought certain crops to Africa, and probably occurring long before that, too. It has been suggested that the wide stone walls of ancient towns and villages in Zimbabwe, like those at the Great Zimbabwe ruins, were designed to deter elephants from raiding crops. So, there's no stopping us using similar deterrents now – building defences to keep elephants away from whatever we need to defend, be it crops, water pumps, or houses.

We can grow 'bio-fences' of thick, thorny vegetation that elephants cannot easily pass through – like Palmyra palm, which is native to India and Southeast Asia; Agave, native to the Americas; and cacti. They can all serve to keep elephants at bay. Such bio-fences have add-on benefits, too, providing fruits or harvestable products. Agave species provide edible flowers, sugar-rich food from the leaf sap and fibres, and sisal hemp fibres. (And if you need more convincing, tequila is also made from certain Agave species.) Bio-fences also capture carbon, reduce soil erosion, are resistant to extreme weather, and can act as wind and fire barriers, further protecting the people and other crops they encircle.

They will not offer immediate protection, obviously, requiring several years to grow to sizes that will deter elephants, but as part of a longer-term plan, they offer a low-maintenance, sustainable, and productive solution.

Whilst such bio-fences are growing, other barriers can be employed such as chilli fences or electric fences. Capsaicin in chilli irritates elephants' eyes and very sensitive noses. Old rags covered in a home-made chilli paste – mixed from fresh chilli, tobacco, and grease – can be hung as 'chilli curtains' on any kind of fencing to deter elephants effectively.

Electric fences can also offer a good solution to keeping elephants at bay, but *only* if they are properly designed and maintained. For an account of the limitations of electric fences, read *The Elephant Whisperer* by Lawrence Anthony. His elephants used to take advantage of lapses in power supply, and he even caught them throwing sticks at the fence wires as though testing whether the electricity was on or not. When they figured it out, they had no problem in pushing over trees to cause a short circuit.

Older elephants have even been seen to use the younger members of the herd to test electric fences by pushing them through them. And if the posts between wires do not have electrified outriggers – branches that poke forward toward the elephants – they can easily pull down a fence by pulling out the posts.

Some elephants have learnt that their tusks do not conduct electricity, and so blithely break electric wires with their tusks and then amble through. There are many rather tragic videos you can watch of this online – tragic because you are literally watching human–elephant conflict in action. Given the design and maintenance costs required, electric fences are often not the best solution for small-scale, subsistence-level farmers.

But one barrier method that is growing rapidly in popularity for small-scale farmers across Africa is beehive fences. Forget mice; giant, mighty elephants are instead afraid of the humble bee. To be fair, African honeybees are not to be underestimated. They are very aggressive, and the threat of being stung by a swarm of angry bees on the trunk, thin skin of the ears or genitals, or around the eyes, is enough to make elephants turn and flee.

The 'Elephants and Bees Project' supported by Save the Elephants shows that hanging purpose-built wooden hives on wire that links all around the crop or settlement to be protected is very effective at keeping elephants at bay. By nudging the wires as they try to enter a field, elephants disturb the hive and anger the bees – and elephants run away at the sound of angry bees. Their smell is a warning enough, so elephants avoid areas that have been sprayed with bee pheromones.

Keeping honeybees also provides an additional revenue source for the farmers, with 'elephant-friendly honey' proving to be very popular with local consumers. And of course, any profit can be used to maintain the beehive fences, providing a sustainable, self-sufficient solution.

People can also hang trip alarms around farms – essentially cowbells strung around on wires – to alert them when elephants are trying to enter the land. This allows people to run out and enact a range of different deterrents, using loud noises (whipcracking, banging drums, firecrackers) or fire – including old rags dipped in kerosene and tied to the end of an aluminium chain that can be lit and swung around, creating the appearance of a fast-moving fireball, or setting fire to homemade 'chilli bombs' that release noxious chilli smoke.

This feels like a good place to stop and reflect for a moment. You heard me right; to defend their homes and maintain their food supplies, people have to go outside in the middle of the night, in darkness, to set off chilli bombs, bang drums, and wave fireballs around their heads. It's worth remembering that, the next time you get stressed waiting in the supermarket queue.

As alien and frightening as these defences may sound to us, they are important, and they do make real, tangible differences to people's lives. The Boteti river forms the western boundary of the Makgadikgadi Pans National Park in Botswana, and people here tend to locate their fields close to the river, for ease of irrigation, with their houses being some distance away. This means elephants and hippos can easily cross over the river, to the farmland on the other side. In this arid area, subsistence-level farming was already tough, even before the elephants returned to Makgadikgadi, but there are times now when it can feel almost impossible.

The charity Elephants for Africa (EfA) are helping people enact a range of different defences against crop-raiding elephants and hippos, and one of their solutions has been to offer relatively low-cost tents, which allow people to sleep within their fields, so they can better protect their crops from any wildlife that may enter at night. The mere presence of being in the field may well be a deterrent, but it also ensures that they can tend to the mitigation tools they have, such as smoking dung and chilli bricks during the night.

One woman, Balongo, took up this solution with gusto. Over the course of one season, she slept in the tent in her field every night except for Easter, meaning she was able to react quickly any time she heard crop raiders approaching. Her crop yield increased so much, from a level where she had barely been able to feed herself and her children, to a point where she had enough money to build a storehouse in which to keep her extra crops.

The storehouse duly was constructed. But when the EfA team went to look at the storehouse, they were immediately concerned – there were hardly any crops inside. Worried, they asked what had happened. Had rats got in? Mould? Vervet monkeys? What had gone wrong?

Nothing. Quite the opposite, in fact. Word had spread that Balongo had extra food, and her storeroom had become a makeshift shop. Over the course of a couple of years, through her own dedication and hard work, she was able to grow and sell enough food to pay off all her debts, including to her children's school for fees stretching back several years, and even their fees for a year in advance, ensuring that her daughters can continue with their education.

Similar success stories of simple mitigation methods changing the lives of individuals or communities can be found across elephant range states – usually wherever there are charities working together with local communities. The range of mitigation methods being tested and used is growing all the time, from simple things like WhatsApp groups to warn farmers across a spread-out area that elephants are approaching, to the high-tech use of drones to scare off elephants, which has recently been trialled in the Tarangire–Manyara and Serengeti areas of Tanzania.

But elephants do not make this work easy. They are very smart, and their cognitive abilities exacerbate the problem, as they can readily learn ways around many of our best defences. They may quickly realise that whilst they are unpleasant, loud noises, fireballs, and the smell of bees aren't going to hurt them. They can learn where the weak spots in a fence are, or which farms are less actively defended, and they may even learn which people are really worth avoiding, and target those who are a bit less aggressive. So successful long-term mitigation against elephant damage requires considerable planning and strategising, and the combined use of a suite of different solutions, now called 'intelligent farming'.

Intelligent farming begins with good land-use planning. The best start is grouping fields and resources together in a concentrated area, where the burden of guarding can be shared, and no one field is isolated and therefore especially vulnerable. Ideally, these areas need to be slightly removed from natural resources the elephants need to access, like water. And although access to the fields needs to be blocked, corridors to essential resources must be left open, because if an elephant is desperate to access water, no amount of fencing, drumming, or chilli is going to stop it!

The community area can be protected with a variety of defences – preferably more than one – including trenches (although these are only viable in areas not prone to flooding, as otherwise elephants could swim across), bio-fences, or chilli fences. A buffer zone of clear, open land or low-growing crops that elephants find unpalatable, such as onions, placed between natural savannah habitat and the community farm is a good idea, as the further elephants have to travel in the open, the more reticent they become – and it allows guards more time to see the elephants approaching. It also helps to define the permissible range boundary for the elephants, helping them learn where is and is not safe to range.

These local-scale defences are important to individuals, but are also crucial for the long-term success of bigger conservation projects. In 2004, the Gregory C. Carr Foundation signed an agreement with the government of Mozambique to work together on the Gorongosa Restoration Project. This 'collaborative management partnership' has been hugely successful in rebuilding and protecting Gorongosa National Park; trees have been planted and bird and mammal populations are recovering. Tourism revenue is increasing, while education, health, reforestation, and employment schemes are having real impacts in the surrounding communities.

Yet this fantastic progress is being threatened by the very elephants the park protects. The rebounding elephant population, a huge success story of Gorongosa, could sour relations dramatically, as some of the animals raid crops of people living in the buffer zones around the park. The principles of intelligent farming are now being employed to try to limit these conflicts, and make sure that elephants do not unknowingly undo all the good work that has been done for conservation and local communities. Importantly, plans are afoot to expand the extent of the park by opening corridors that will eventually stretch all the way to the coast.

In fact, with its anti-poaching patrols, conflict mitigation schemes, balanced land-use planning, community development programmes (which include programmes aimed specifically at educating girls and involving women in conservation) and plans to open up wildlife corridors, Gorongosa represents practically the best of everything we can do to secure a place for elephants in the world. It is the kind of project that can only inspire hope, showing us how nature can be restored and conserved, and used to develop and uplift communities sustainably.

Arguably the most vital element in the success of Gorongosa so far lies in the relationship the park and restoration project is fostering with the people. Local communities are involved as key stakeholders, including becoming rangers in the park; Mozambican scientists and managers hold key leadership positions; and the role of women is being especially promoted, as the need for their input is increasingly being recognised. Giving women a seat at the negotiating table during peace talks attempting to end armed conflicts results in much better outcomes – they are 64 per cent less likely to fail, and peace is 35 per cent more likely to still be holding fifteen years later! That's a pretty staggering improvement.

In general, higher levels of gender equality are associated with lower conflict rates both within and between states. Likewise, educating girls, promoting women's health issues, and involving them in planning and implementation is only going to improve conservation prospects.

It is not that women have all the answers necessarily – not all the time, anyway – but that women often bring a different perspective, due to having different kinds of interaction with elephants. After all, it is often women or children in Africa who collect the water for households, and water sources can be a prime source of conflict with elephants. Taking account of all sides and hearing all voices should ultimately result in more comprehensive and viable plans.

To give us even more reason to be optimistic, Gorongosa is not alone in these successes. Other collaborative management partnerships are showing equally positive signs of impact, across Africa. The Gonarezhou Conservation Trust in Zimbabwe, a partnership between Frankfurt Zoological Society and Zimbabwe Parks and Wildlife Management Authority, is successfully protecting the region's elephants and rejuvenating the area. And as part of the Greater Limpopo Transfrontier Conservation Area mega-park, securing Gonarezhou (which means 'place of elephants' in the local Shona language) is particularly important, as it will allow safe movement of elephants over vast areas between South Africa, Mozambique and Zimbabwe.

African Parks, of which HRH Prince Harry, the Duke of Sussex, is the president, is another organisation that has set up partnerships to manage numerous protected areas across Africa. These are now offering safe habitat for elephants in Benin, Central African Republic, Chad, Republic of Congo, DRC, Malawi, and Rwanda – many of the areas worst hit by poaching, in fact, as well as education, employment, and tourism revenue for local communities.

It is not only large organisations that are making important differences. Low-budget projects are also achieving a huge amount in localised areas. Staff at the Mali Elephant Project, which works to protect the remnant desert-adapted elephant population in the country's Gourma region, realised that although local people were concerned by elephants crop raiding or blocking their entrance to forests, they overwhelmingly wanted to conserve them. People were worried that losing elephants must mean their environment was degraded, and they did not want that to be the case.

The Project works with local communities, encouraging and empowering them to devise their own sustainable-resource management systems based on their local cultural knowledge, which allows natural resources to be shared between people and elephants, not competed over. Not a single elephant has been lost to poaching in the region for two years, and households participating in trials of their livelihood schemes – which involve actions such as marketing wild fruits and medicinal plants, and rearing only a few fat cattle instead of a large herd of thin ones – resulted in an average 400-per-cent increase in household income! That is a truly momentous achievement.

In the northern tip of the remote and rugged Matthews Mountain Range of Kenya, not far from the turquoise wonder of Lake Turkana, the Milgis Trust is likewise promoting peaceful coexistence between elephants and the local Samburu people. This is a region entirely outside of protected areas and with no fencing in sight. By employing community scouts, building schools, funding clinics, and installing boreholes to access water, and mostly by educating people about elephants, tolerance has increased and fear decreased. The number of elephants in the area is on the rise, while poaching and conflict have decreased substantially.

Importantly, the community view these increasing elephant numbers as a blessing. Education, development, and social participation have been key, yet again. Of course, it is not only rural communities who can have problems with elephants. Many reserve managers – especially in fenced or comparatively small reserves – are concerned about elephants damaging trees, particularly iconic baobabs or marula trees, for example, or overfeeding on vegetation and so degrading the habitat. Many solutions in South Africa are being adopted to protect individual trees – hanging beehive fences, or packing sharp rocks at the base of trees to prevent elephants from approaching them too closely. The best solution for any kind of damage or overuse is usually to allow dispersal of elephants by opening up corridors and encouraging mega-parks. Now is the time to encourage the opening of corridors and linking of parks like never before, to give elephants the ability to move. Away from us. Improving the connectivity of elephant habitat can be achieved in some surprising ways. In Kenya, underpasses are increasingly being used to allow elephants – and all other wildlife – to move safely under roads or railway tracks that have been built through their habitats.

Around Mount Kenya, a busy highway had effectively split the habitat for many years, and left elephants and motorists vulnerable to high-speed collisions. An underpass was built and completed in late 2013, with the first pioneer young bulls using it to cross under the road a few weeks later. It is now used routinely by many males and family groups, reconnecting the Mount Kenya elephant population that had formerly been split – about 2,000 individuals had been stuck on the mountain above the road, and 5,000 in the forests and plains below the road.

Similar underpasses have also been incorporated into the new railway line that connects Nairobi with Mombasa, which runs through Tsavo East National Park. It is hoped that these underpasses will prevent or limit wildlife deaths from train collisions in the park, and it's a good example of how a bit of creative thinking and awareness at the design and planning stages of our ever-expanding infrastructure projects can help reduce the environmental threats.

Funding corridors and underpasses is not straightforward, and often relies on external donors. Corridors can be especially tricky, as large tracts of land may need to be purchased or leased. But leasing degraded lands from communities and restoring them to natural habitats for wildlife ticks so many boxes that we cannot simply file the idea away under the 'too difficult and complicated' category. One important wildlife corridor that radiates west out of Amboseli National Park is kept open because conservation agencies pay an annual fee to the landowners. Alternatively, landowners have become inclined to convert land to corridors themselves, when such areas could act as conservancies that attract tourists.

Farmers could even be encouraged to share important dispersal lands with elephants – whilst maintaining their existing farming activities in a more concentrated area – with direct payments. Whereas compensation schemes only pay out if damage has occurred, here farmers are rewarded for sharing land, allowing elephants and wildlife to pass through it. The guaranteed income could offset any crop losses the farmers may endure, and allow them to invest in well-designed mitigation methods.

But yes, all of this needs money. So far, conservation finance in many elephant range states relies heavily on tourism and cash injections from rich philanthropists or charitable organisations. As you have probably noticed, many if not all of the most successful projects promoting coexistence are being planned, funded and implemented by non-governmental charities, and often at very small, local scales. It is time governments stepped up to implement these schemes on national scales across the continent.

We need international governments to talk sensibly about what is needed and what each can contribute to support elephant range states tangibly and practically in meeting their conservation and development needs. The unique natural heritage of Africa must be included in discussions of socio-economic development.

Economic growth, sustainable development, and nature conservation are not conflicting aims. They must be linked, and education and social stability are central to achieving this. People around the world should be encouraged to understand the importance of nature; and improved governance, decreased corruption, and better education can give people the support to remain tolerant even of species that can be problematic at times.

But ultimately, it will all come down to money. Which means that political will is essential.

Habitat that supports elephants could be viewed as a 'global public good', given that the entire world benefits from its continued existence. International levies, whereby countries contribute a fixed percentage of funding to relieve some of the financial burden put on individual elephant range states and non-government organisations, is therefore one funding avenue. The reality of this kind of plan may feel like a long way off, but similar schemes were designed as part of the Paris Climate Agreement, to help fund the implementation of climate-change initiatives in developing nations.

Extending such schemes to cover nature conservation and biodiversity protection should not be beyond reach, especially as securing biodiversity contributes significantly to carbon capture and achieving climate targets. Nature – and elephants – are necessary for a healthy world.

What it all requires, to ensure *we* can live peacefully alongside elephants from now on, are changes made by *us*. We must adapt first, make concessions, and create solutions that allow elephants to fit in again. People are key, but in all this planning and discussion, we must never lose sight of the elephants – of their unique biology, their social and spatial requirements, and their long lives lived among family and friends.

We must respect these relationships and make sure that all our management and conservation schemes of the future do not fracture their associations. We must learn from our mistakes and not repeat the wrongs of the past. Elephants need space to move and associate with who they want, when they want. That is something we must accept and learn to live with if elephants are to survive.

None of the necessary actions are easy, they may require difficult decisions and innovative solutions, but globally we are making more and more pro-environmental decisions, and there are many reasons to be optimistic. The elephant losses we have witnessed can be halted, conflict can be eased, and coexistence can be achieved. Corridors and mega-parks can be secured, facilitating natural dispersal and balance and allowing elephant numbers to recover; and people living alongside elephants can be raised out of poverty, valuing the nature they live alongside.

I say all this can be done, because I am hopeful. But if it doesn't happen, the last giants will become extinct in the very near future.

We owe it to them, and ourselves, not to let this happen.

Epilogue

What can we do, sitting here, far removed from the wildlife reserves and plains of Africa, to help achieve all of this? One obvious answer is to make a financial contribution to any of the numerous, excellent elephant conservation charities that work to change lives on the ground. If you can't give money, then time and expertise is often as gratefully received.

Many of these organisations were set up by field biologists working in Africa, who have no marketing expertise or accountancy skills, or understanding of the law, or logistics, or good administration. Offering a few hours of your time to help file their records or update their websites could be unimaginably useful.

If you can, get out there and visit these countries as responsible tourists. Pay park and reserve fees, employ local guides and drivers, visit education programmes and community conservation projects, buy elephant-friendly honey or elephant-dung paper produced by local development projects.

Putting pressure on our politicians and policymakers to take environmentally sound decisions remains critically important to ensure that economic growth is not promoted at the cost of natural places; that habitat fragmentation and pollution are properly dealt with and not ignored; and to make sure that ivory stays on elephants, not our mantelpieces. Civil society has the power. Use your voice, as elephants don't have one.

Most, and simplest of all – remember to enjoy, respect and value nature.

Timeline

When	Geological time[*]	What happened
~100 million years ago (mya)	Cretaceous period	Afrotheria lineage, which includes elephants, splits from other mammalian lineages, including the branch that led to humans and other primates. This is the date of the last common ancestor between humans and elephants.
65mya	Cretaceous period	Extinction of the dinosaurs.
~65–55mya	*Palaeocene* epoch of Paleogene period	Evolution of **first proboscideans**, when the lineage that resulted in elephants split from that of the hyraxes and manatees. These early proboscideans were all extinct by ~30mya.
~30–25mya	Late *Oligocene*	Evolution of **deinotheres** in Africa, which subsequently colonised southern Asia and Europe. The deinothere group died out ~1mya.
~25mya	Late *Oligocene*	Evolution of **mastodons** in Africa, which colonised most of the northern hemisphere. The last of the mastodons died out around 10,000 years ago.
~24mya	Late *Oligocene* or early *Miocene*	Evolution of **gomphotheres**, which colonised North, Central, and South America, and Eurasia. Extinction less than 10,000 years ago.
~12–10mya	*Miocene*	Evolution of **elephantoids**, the super-family that contains today's elephant species.
~8–6mya	Late *Miocene*	Last common ancestor between ***Loxodonta*** lineage and Asian elephant/mammoth lineage.
~7–6mya	Late *Miocene*	Last common ancestor between humans and our closest living relative, the chimpanzees.
~7–5.5mya	Late *Miocene*	Mammoth and Asian elephant lineages split.
~5.5–5mya	Late *Miocene*	***Loxodonta africana*** (savannah elephant) and ***Loxodonta cyclotis*** (forest elephant) lineages split.

[*] Plain text denotes geological 'eras', some of which are typically broken into shorter 'epochs', which are named in italics.

~5mya	Start of *Pliocene*	Earth cools and becomes drier, forest cover shrinks.
~3mya	End of *Pliocene*	Migration of ***Elephas*** species from Africa to Asia.
~2mya	Start of *Pleistocene* epoch of Quaternary period	Rapid cooling of Earth.
		Appearance of human lineage, ***Homo***.
~500,000 years ago	*Pleistocene*	Clear evidence from stone tools that ***Homo*** species were hunting proboscids.
~195,000 years ago	*Pleistocene*	Earliest known fossils of our own species – anatomically modern humans – ***Homo sapiens***.
50–30,000 years ago	*Pleistocene*	Extinction of ***Palaeoloxodons*** (straight-tusked elephants) from mainland Europe and Britain.
15–10,000 years ago	Late *Pleistocene* to early *Holocene*	Extinction of most proboscids across the Americas, Europe and Asia. ***Homo sapiens*** settles most parts of the planet.
~6,000 years ago	*Holocene*	Extinction of last **gomphotheres** in South America, and last ***Palaeoloxodons*** in Greek islands.
~5,000 years ago (3000 BC)	*Holocene*	First evidence of Asian elephants being captured and trained in India.
~4,000 years ago (2000 BC)	*Holocene*	Extinction of last woolly mammoths from Wrangel Island in Siberia.
~2,000 years ago (218 BC)	*Holocene*	Hannibal crosses the Alps with an army of African elephants.
~1,500 years ago (AD 500)	*Holocene*	The last elephants disappear in North Africa.
~500–100 years ago (1500–1920)	*Holocene*	Expansion of agriculture in Africa; European colonisation and ivory trade grows; the demise of Africa's elephants begins.
1920–2019	*Holocene*	Massive boom in human population. Elephant numbers drop from 12 million to less than 415,000.

Further Reading

Much of the information and data presented in this book has been taken directly from the original scientific papers, which, due to the vagaries of academic publishing, are not always easily accessible to anyone outside of a research institution. So rather than listing every source, which you may not be able to access, I have provided details of the research leaders you might want to look up, and also list many book and web-based resources where it is possible to obtain further information about the research and the issues we discussed here.

The Amboseli elephants of Kenya are undoubtedly the best studied elephant population in the world, having been observed continuously since 1972 by the legendary Cynthia Moss, and other members of the Amboseli Elephant Research Project (the research wing of **The Amboseli Trust for Elephants**). The team of researchers at ATE includes Phyllis Lee, Joyce Poole, Norah Njiraini and Katito Sayialel, and they regularly collaborate with Beth Archie, Richard Byrne, Patrick Chiyo, and Karen McComb, as well as Lucy Bates and Graeme Shannon.

So much of the knowledge we have about African elephant behaviour has its roots in the Amboseli study, which continues to this day. You can learn more about the work of Cynthia and the Amboseli team from their website: www.elephanttrust.org, and if you are especially keen to read in detail about the vast information they have amassed, it is all laid out in their book: *The Amboseli Elephants: A Long-Term Perspective on a Long-Lived Mammal* by C.J. Moss, H. Croze and P.C. Lee, Chicago University Press, 2011.

The other foundational research project on savannah elephants is, of course, **Save the Elephants** (www.savetheelephants.org), established by Iain Douglas-Hamilton and George Wittemyer and based in the Samburu and Buffalo Springs National Reserve of Kenya. Iain was the pioneer of elephant field-study, as described in his early books *Among the Elephants* and *Battle for the Elephants*.

The current data on elephant numbers and status is taken from the results of the Great Elephant Census, conducted by Dr Mike Chase of **Elephants Without Borders**, and the African Elephant Status Report of the IUCN African Elephant Specialist Group. Details can be found here: http://www.greatelephant-census.com and https://www.iucn.org/ssc-groups/mammals/specialist-groups-a-e/african-elephant.

There is a large number of charities working to advance elephant conservation. Those from which we borrowed data, examples or information, are:

African Parks
Big Life Foundation
Ecoexist
Elephants Alive
Elephants for Africa
Elephant-Human Relations Aid
Elephant Voices
Elephants Without Borders
David Shepherd Wildlife Foundation
David Sheldrick Wildlife Trust
STE Elephants and Bees
Mali Elephant Project
Mara Elephant Project
Mount Kenya Trust
Save the Elephants
Space for Giants
Tsavo Trust
The Tusk Trust

Information on specific topics can be found at the following resources:

To learn more about elephant natural history, try:

Clive A. Spinage, *Elephants*, Poyser Natural History, 2003.
Raman Sukumar, *The Living Elephants*, Oxford University Press, 2003.

Elephant communication has been studied in detail by:

Dr Joyce Poole: http://www.elephantvoices.org.

Dr Caitlin O'Connell in Etosha: http://utopiascientific.org/ Research/index.html.

Dr Beth Mortimer at Samburu: https://www.savetheelephants. org/category/publications/.

Joyce and Caitlin have also added a lot of information about male elephants, as has Dr Kate Evans at: http://www. elephantsforafrica.org/research/behavioural-ecology/.

Dr Shifra Goldenberg: https://www.savetheelephants.org/category/publications/.

Dr Michelle Henley: http://elephantsalive.org/.

Dr Hannah Mumby: https://www.hannahsmumby.co.uk/research.

Josh Plotnik conducted the studies on cognition and mirror self-recognition in Asian elephants: http://thinkelephants.org.

John Allman has investigated von Economo neurons: https:// www.smithsonianmag.com/science-nature/ brain-cells-for-socializing-133855450/.

Genetic analysis of desert elephants in Namibia has been conducted by Yasuko Ishida and Alfred Roca. For a summary, see: https:// www.sciencedaily.com/releases/2016/08/160803161607.htm.

Mark Deeble's blog, 'A wildlife filmmaker in Africa', and film *The Elephant Queen* give more details about Satao, the large tusker of Tsavo.

If you would like to know more about intelligence, cognition, and learning in animals generally, you could do much worse than reading:

Brian Hare and Vanessa Woods, *The Genius of Dogs*, Oneworld Publications, 2013.
Frans de Waal, *Are We Smart Enough to Know How Smart Animals Are?*, Granta Books, 2016.

For discussions of what it might be that sets us humans apart from other animals, try:

Kevin Laland, *Darwin's Unfinished Symphony: How Culture Made the Human Mind*, Princeton University Press, 2017.
Michael Tomasello, *Becoming Human*, Belknap Press, 2019.

For more information about the elephants of Makgadikgadi: http://www.elephantsforafrica.org.

For information about pathfinder young males: http://elephant-salive.org/.

For information about the Tarangire elephant population in Tanzania: https://tanzania.wcs.org/landscapes/tarangire-ecosystem.aspx.

For information about Addo and Kruger National Parks: https://www.sanparks.org.

For information about Gorongosa National Park in Mozambique: www.gorongosa.org.

For information about forest elephants, as studied by Andrea Turkalo at Dazanga and Fiona Maisels: https://blog.wcs.org/ photo/author/aturkalo/ and https://www.savetheelephants. org/category/publications/.

The work of the **Amarula Elephant Research Programme**, led by Rob Slotow at the University of Kwa Zulu Natal, has contributed much to our knowledge of elephant ecology, sustainable hunting rates, and the behaviour of juveniles raised as orphans in South Africa. To read more about the delinquent males of Pilanesberg specifically, see: www.movinggiants.org/ stories/the-delinquents.

Extensive work on elephant ecology and role of mega-parks has been conducted by Rudi van Aarde at the Conservation Ecology Research Unit of the University of Pretoria.

The research looking at ivory prices and demand has been conducted by, among others: Colin Beale; Severin Hauensteing; Monique Sosnowski; and George Wittemyer. Sam Wasser has done much of the work determining the origin population of ivory, and showing that ivory shipments are being moved by organised criminal syndicates.

Studies on the social effects of poaching have additionally been conducted by Kathleen Gobush, and Hamisi Mutinda.

Further information about the illegal ivory trade and its implications for elephants can be found in articles on the *National Geographic* website, particularly in articles by journalists such as Bryan Christie, Dina Fine Maron, and Rachel Nuwer, as well on organisational websites such as TRAFFIC and CITES. **Save the Elephants** has commissioned a lot of research into the ivory trade, particularly conducted by Esmond Martin and Lucy Vigne. Check out the **Save the Elephants** and **WildAid** websites for more details.

Discussions about trade – and whether it should be legalised – can be deeply polarising. Articles by researchers such as Katarzyna Nowak and Ross Harvey on *The Conversation* website give good overviews of the pitfalls of pro-trade arguments, and they link to alternative viewpoints by pro-trade proponents.

Just like debates about legalising ivory trade, opinions about trophy hunting are easy to come by on the internet, though much of the content may be short on science and heavily biased one way or another. For relatively balanced views, again search on *National Geographic*, *The Conversation*, and *Africa Geographic* websites. Pieces by authors such as Katarzyna Nowak, Ross Harvey, and Simon Epsley give good background, although it is evident that they all come down on the side against elephant trophy hunting. An article by Michael Paterniti in *National Geographic* titled 'Trophy Hunting: Should We Kill Animals to Save Them?' gives a good overview of the general arguments for and against.

Finally, there are a number of anecdotal accounts from conservationists about their own relationships with elephants. Some of my own favourites include:

Lawrence Anthony, *The Elephant Whisperer*, Sidgwick & Jackson, 2009.

Iain & Oria Douglas-Hamilton, *Among the Elephants*, BCA, 1975.

Katy Payne, *Silent Thunder*, Penguin, 1999.

Sharon Pincott, *Elephant Dawn*, Allen & Unwin, 2017.

Don Pinnock & Colin Bell, *The Last Elephants*, Hardie Grant Books, 2019.

Acknowledgements

This book was very much a collaborative effort. I have already mentioned the invaluable contribution made by Dr Lucy Bates and Dr Graeme Shannon and have listed many of the organisations that have helped me to research this book. I think it's clear to the reader that such a vast topic as the history, both ancient and modern, of an entire specie, as well as a summary of elephants' biology, psychology and place in ecology requires an enormous amount of background reading.

There are many people that I must thank for their advice, help and assistance along the way. This journey really began way back in the early 1990s, when my father took mc to meet the late David Shepherd, and it's to him that I owe my fascination with elephants.

The planning for such a mammoth task (excuse the pun) took many years of research, field trips and expeditions and I would especially like to thank Charlie Mayhew and Mary-Jane Attwood at the Tusk Trust, who have very generously put me in touch with many of their contacts throughout Africa. I'd also like to thank the team at Virunga National Park, Mike Chase at Elephants Without Borders and the staff at the David Sheldrick Wildlife Trust in Kenya for their hospitality, as well as Nigel Winser and Shane Winser at the Royal Geographical Society.

In addition, I'd like to thank the team at October Films for producing such a wonderful series about the elephant migration in Botswana for Channel 4 and Animal Planet, especially Alexis Giradet for being such a good director and entertaining company. My utmost appreciation as well goes to all those who have accompanied me on my elephant journeys over the years: Dave Luke, Neil Bonner, Simon Buxton, Will Charlton, Alberto Caceres, Chris Mahoney, Hardus Vermaak, Parker Brown, Beki Henderson, Mike Holding, Tania Jenkins, Gareth Flemix, Kane Motswana, Boston Ndoole, Max Graham, Mike McCartney, Ruthie Markus, Pete Meredith, Dave Southwood, Richard Harvey, Graham McCulloch, Anna Songhurst, my brother Pete, and my parents Levison and Janice Wood.

I am indebted to all those that helped with the research and editorial process of the book: Kate Harrison, Charlotte Tottenham, Ash Bhardwaj, and Tash Turgoose for her fantastic illustrations.

I am grateful too for the support of all those companies and organisations that have helped out over the years: Global Rescue, The Belmond Group, Nomad Travel, Taylor Morris, Leica Cameras, Belstaff, IWC Watches and Oliver Sweeney.

As ever, I owe the book to my fabulous agent Jo Cantello, and Rupert Lancaster at my publisher, Hodder & Stoughton, as well as all of the team involved, especially Rebecca Mundy, Cameron Myers and Barry Johnston.

Finally, my gratitude to the people of Africa for their generosity and kindness, and in particular the inspirational individuals and teams who work in the world of conservation to help protect not only elephants, but all species.

Picture Acknowledgements

© British Library Board. All Rights Reserved / Bridgeman Images: 2, middle and bottom.

© CARL DE SOUZA / AFP via Getty Images: 13, bottom.

© Earl Theisen / Getty Images: 10, bottom.

Photograph by Eliot Elisofon, 1972, EEPA EECL 10138 © Eliot Elisofon Photographic Archives / National Museum of African Art / Smithsonian Institution: 3, bottom.

© Graeme Shannon: 6, top. 8, top. 9, bottom. 11, top.

© Granger Historical Picture Archive / Alamy Stock Photo

© Hulton-Deutsch Collection / CORBIS / Corbis via Getty Images

© Shutterstock.com: 1, top. 2, top. 3, top left. 3, top right.

© Luca Galuzzi www.galuzzi.it CC BY-SA 2.5: 1, bottom.

From The New York Public Library NYPL: b11721700: 10, top.

© Peter6876 / Dreamstime.com: 12, top.

© Simon Buxton: 16, bottom.

All other images © Levison Wood

www.tusk.org

Tusk's mission is to amplify the impact of progressive conservation initiatives across Africa.

We partner with the most effective local organisations, investing in their in-depth knowledge and expertise. By supporting and nurturing their conservation programmes, we help accelerate growth from an innovative idea to a scalable solution.

Since 1990, we have helped pioneer an impressive range of successful conservation initiatives across more than 20 countries, increasing vital protection for over 10 million acres of land and more than 40 different threatened species. But the threat to Africa's unique natural heritage remains real and more urgent than ever.

With your generous support, Tusk can, and will, continue to have a positive impact in Africa.

How You Can Help
Find out more and support our work at www.tusk.org

Tusk Trust Ltd, 4 Cheapside House, High Street, Gillingham, Dorset SP8 4AA
Tel: +44 1747 831005   Email: info@tusk.org
UK Registered Charity No 1186533
Website: www.tusk.org;   Facebook: www.facebook.com/tusktrust   Twitter: @tusk_org